Great Mambo Chicken and the

Transhuman Condition

Great Mambo Chicken a

e Transhuman Condition

Science Slightly Over the Edge

ED REGIS

Addison-Wesley Publishing Company, Inc.
Reading, Massachusetts   Menlo Park, California   New York
Don Mills, Ontario   Wokingham, England   Amsterdam   Bonn
Sydney   Singapore   Tokyo   Madrid   San Juan

Acknowledgment of permissions granted to reprint previously published material appears on page 301.

Library of Congress Cataloging-in-Publication Data

Regis, Edward, 1944–
    Great mambo chicken and the transhuman condition : science
slightly over the edge / Ed Regis.
        p.   cm.
    Includes bibliographical references.
    ISBN 0-201-09258-1
    1. Science—Miscellanea.   2. Engineering—Miscellanea.
3. Forecasting—Miscellanea.   I. Title.
Q173.R44   1990
500—dc20   90-382
                                                                      CIP

Jacket design by Gary Koepke
Text design by Joyce C. Weston
Set in 11-point Galliard by DEKR Corporation, Woburn, MA

  BCDEFGHIJ-MW-943210
Second printing, October 1990

For William Patrick

Contents

The Mania

When Saul Kent had the suspension team at the Alcor Life Extension Foundation in Riverside, California, surgically remove the head of Dora Kent, Saul's mother, from her body, his hope was that she could eventually be restored to life and health, probably even youth. The *last* thing on his mind was that they'd all wind up being investigated for murder.

Murder! The thought was entirely ludicrous—not, indeed, that any of them had given even a moment's consideration to that possibility at the time of the event. After all, the two classical signs of life, respiration and heartbeat, had vanished minutes before, and so Dora Kent, at age eighty-three, was for all intents and purposes now legally dead. The hope of the surgical team was that at some point in the distant future a fresh, new body could be cloned for Dora Kent from one of her old cells. Her old brain would then be placed inside the head of the new body, after which her brain would be revived and the patient would come back to life just as if she had been awakened from a very long sleep. Once she was "reanimated," in the cryonics jargon, Dora Kent would enter her "second life cycle" (also part of the jargon), and go on to lead a long and prosperous new life. A very *long* life, perhaps: she might live for hundreds, maybe thousands, of years. She might even become immortal.

Everyone present knew that her eventual resurrection was a long shot, to say the least. Between her death in 1987 and her hoped-for revival at some indefinite time in the future, Dora Kent's head would remain frozen in a tank of liquid nitrogen, at a temperature of −186°C, and no one in the cryonics business was expecting that when defrosting time came the patient would just magically spring back to life. Cryonic suspension was clearly a last-ditch measure. "No one wants to be frozen," Saul Kent once said. "Being frozen is the second-worst thing that can happen to a person. The only thing worse is dying *without* being frozen."

Nevertheless, it wasn't as if they were expecting men from Mars to revive Dora Kent. The fact was that their expectations rested on nothing more than the normal and ordinary progress of science, just plain science, the plain, old-fashioned, everyday science and technology that had already accomplished so many stupendous feats, things that only a few years previously were regarded as "impossible": the moon landings, the heart transplants, the gene splicings, and all the other modern miracles. In the waning days of the twentieth century it did not take an especially gigantic leap of the imagination to think that at some point in the future it would be entirely possible to thaw out a frozen brain, implant it in a new body, shock it into conscious awareness, and restore it to normal functioning. In fact, the members of the suspension team had some rather specific notions about the precise science that would be required to accomplish the task. It was called "nanotechnology," and was already on the conceptual horizon.

Nanotechnology had been invented back in the 1970s by an MIT grad student by the name of Eric Drexler. As Drexler conceived of it, his invention, when it was perfected (thus far it was still in the idea stage), would give you complete control over the structure of matter. It would make possible the direct manipulation of matter at the atomic level—"atom by atom," as he described it. This would be accomplished by an army of robots, each of which was roughly the size of an individual molecule.

A big enough collection of these tiny robots would be able to do anything that was capable of being done with matter. They'd be able to take molecules of ordinary carbon—charcoal, for example—and reassemble them in the form of diamond crystals. They'd be able to take inanimate raw materials and fashion them into living organisms, creating new life from scratch. They'd be able to repair damaged biological cells one by one. In fact, when Drexler had first come up with the idea of these miniature robots back in 1976, one of the first applications he'd then thought of was frostbite treatment. You could send phalanxes of these machines coursing through the bloodstream to locate the ailing cells, find out what was wrong with them, and make the necessary corrections.

It sounded miraculous even to Drexler, but he kept on reminding himself that this is exactly what biological cells did on a regular basis, every day of the week: they automatically maintained their own metabolic equilibrium, they repaired themselves and each other, they gave rise to fresh, new cells—and what's more they did all these things *on their own,* without any intelligent supervision whatsoever. And so, for example, if the white blood cells—which lacked any "brains"—had evolved the ability to scavenge and destroy threatening bacteria from the bloodstream, then there was no reason why even more complex biological repair capabilities couldn't be intentionally programmed into the molecular robots he was imagining.

The application to cryonics was then obvious. In the process of cryonic suspension, many of the body's cells underwent freezing damage and would suffer further injury in the process of thawing out. But as long as the underlying cellular information structure had been left reasonably intact, then a fleet of molecule-sized robots could get inside the ailing cells, diagnose their troubles, and put them right again. At length Drexler's little nanotechnological marvels would have returned the frozen body back to where it was before it ever died, good as new.

Anyway, when officials of the Alcor Life Extension Foundation were under threat of a murder charge, their attorney in Los Angeles, Christopher Ashworth, contacted Eric Drexler and some other forward-looking scientists to get together a batch of depositions, or "technical declarations," as they were called, that could be used in court to make cryonics seem less of a bizarre freak show of science, which was how lots of people, including many scientists themselves, then regarded it. Drexler was happy to oblige. He was then a visiting scholar at Stanford University, where he was teaching the world's first college course on nanotechnology.

"Future medicine," Drexler said in his technical declaration, "will one day be able to build cells, tissues, and organs and to repair damaged tissues. This, obviously, would include brain tissues suffering from preexisting disease and the anticipated effects of freezing. These sorts of advances in technology will enable patients to return to complete health from conditions that have traditionally been regarded as nonliving, and beyond hope, i.e., dead."

So here was a Stanford University scientist—which is to say, a researcher at one of the world's greatest universities, known to be particularly strong in the sciences—here he was saying that future medicine would be able, in effect, to *bring the dead back to life*.

But in this belief, Eric Drexler was not alone. Chris Ashworth had also gotten similar technical declarations from other like-minded scientists at Harvard, Columbia, Johns Hopkins, and elsewhere, all of them supporting the idea that raising the frozen departed was not some lunatic-fringe delusion, but instead a reasonable prospect, well grounded in current fact and likely future advances.

Then there was the declaration of Hans Moravec. Moravec was a roboticist at Carnegie-Mellon University—he was in fact director of the Mobile Robot Laboratory there—and he offered up what was, for an advanced thinker like himself, a rather typical statement. "It requires only a moderately liberal extrapolation of present technical trends," Moravec said in his declaration, "to admit the future possibility of reversing the effects of particular diseases, of aging, and of death, as currently defined."

Privately, though, Moravec was not overly fond of cryonics. "I'm actually not a strong believer in it," he said. "It seems such a crude technology. The thing is, when you resurrect a person who has died of old age, that person is more than half gone already."

Not that he thought resurrecting the dead was impossible. It was *entirely* possible: "There is more scientific justification for this possibility," he'd said in the declaration, "than for the afterlife beliefs of many of the major religions." Moravec's doubts stemmed basically from the fact that he himself had a better idea. In truth, compared to what he *really* thought was going to be possible in the near future—meaning within the next fifty years—cryonics was only a stopgap measure, a bridge between now and the time when all of us would be able to live forever. Moravec's own scheme was downloading.

"Downloading" is a computer science term for taking information out of one computer and transferring it into another. Moravec's plan was to read out the information stored inside a human brain and load it into a computer outside the body. In the end, that person would *become* that computer.

To Moravec, as well as to lots of other theorists, the human brain was no more and no less than an extremely complex biological machine. That being so, it followed that the human personality—your character, your emotional life, your tastes and aspirations—all these were nothing more than programs, software, patterns of information embedded within and among the brain cells. And if *that's* so, well then, Moravec reasoned, what was the obstacle to extracting that information from the biological brain and transferring it into a nonbiological one outside the body? Information is information: if it works in the brain, it'll work just as well on the circuit board. It was simply a matter of going from "wetware" to hardware.

Quite a metamorphosis, by any standard. The old "wetware" anatomy would be put out with the garbage, and the new silicon-chip "person" (or whatever it was) would spring into life and conscious awareness. You'd be losing the body—you'd lose the world, the flesh, the devil—but think of what you'd be gaining: freedom from physical constraints, faster thinking speed, a bigger memory. You could trade data with other downloaded minds, maybe even merge yourself into another person's experiences to become an enlarged and amplified thinking entity.

In fact, it would be a kind of heaven on earth. Think of how throughout the ages all manner of religious ascetics, puritans, and self-flagellants had spent their days bemoaning the flesh and its ills: the body, according to St. Augustine, was "evil, sordid, bespotted, and ulcerous." Think of how they'd longed for blessed release from the worldly travails and disappointments—the vale of tears, and all the rest—of how they'd wanted to transcend physical limitations and go on up into heaven, into the realm of pure spirit, where they could exist in the company of God and the angels.

Naturally, these religious souls had imagined that in order to do this—in order to become pure spirit—you'd first have to *die*. And you couldn't hurry it up, either, because that would be suicide, a mortal sin, and would only get you eternal hellfire and damnation. So you'd just have to wait for your own ulcerous and bespotted body to stop functioning, and *then* you'd be able to go on up to the world of pure spirit and life everlasting. But Hans Moravec had invented the perfect way of doing this *on earth*, at your own choice and option, and—best of all—*without dying!*

Think of it! It would be a temporal, corporal, quite this-worldly way of escaping all the same ills and limitations of the flesh, just exactly as had been envisioned by the greatest saints. It would be a pathway to true immortality, only you'd have it *now*, and *right here on the home planet*.

The only thing you'd have to worry about would be a power outage, but even *that* would be no real problem. You could get around that by storing one or more backup copies of yourself in another computer, or on disk, or in whatever advanced storage media they might be using fifty years from now. For that's when all of this was supposed to happen, in *just fifty years!* And the thing of it was, you could do all of it with *science*, just plain, old-fashioned, ordinary science, no religious-mystical mumbo jumbo about it.

An outsider might wonder about Moravec and his wife, Ella, a born-again Christian going for her master's degree in divinity at a seminary outside Pittsburgh. You might think they were a little bit ill matched, the divinity student and the director of the Mobile Robot Lab, but Moravec himself never thought so. "The idea that your essence is software seems a very small step from the view that your essence is spirit," he said. Meaning that both he and his wife were really, at bottom, interested in the same thing, which is to say, true immortality, life everlasting in the form of pure consciousness. Ella, it's true, never shared her husband's passion for the hard-science way of getting there. She wanted to get to heaven in the traditional way, through prayer, devotion, and good works. But her final goal was essentially the same as her husband's, so the two of them—the roboticist and the divinity student—well, they made the perfect couple.

And to be strictly accurate, Moravec was not aiming for a heaven *on earth* anyway. What he envisioned was a vast interstellar culture, a population of superintelligent robots and disembodied postbiological minds spread out across the stars and the galaxies. Even Eric Drexler, for that matter, wanted to get humanity *off* of the earth, and both he and Moravec had tried to think up ways of getting people up there without relying on the standard chemical rocket technology, which was, after all, as old as ancient China. Thus Drexler had written his master's thesis at MIT on solar sailing and later got two patents on aerospace inventions, while Moravec, for his part, had figured out a way of using rotating space tethers—they'd be like large bolas, truly gigantic rotating catapults—that would fling payloads up into orbit on their own momentum.

What these forward-looking scientists were doing, it turns out, was nothing less than reinventing Man and Nature. They wanted to re-create Creation. They wanted to make human beings immortal—or, failing that, they wanted to convert humans into abstract spirits that were by nature deathless. They wanted to gain complete control over the structure of matter, and they wanted to extend mankind's rightful sovereignty out across the solar system, into the Galaxy, and out into the rest of the cosmos. An imposing enterprise, to be sure, but that was the way of science and technology during these bold days of *fin-de-siècle* hubristic mania.

Fin-de-siècle hubristic mania was the desire for perfect knowledge and total power. The goal was complete omnipotence: the power to remake humanity, earth, the universe at large. If you're tired of the ills of the flesh, then *get rid of the flesh:* we can *do* that now. If the universe isn't good enough for you, then *remake it*, from the ground up.

Hans Moravec once complained about *matter itself* that it wasn't really *doing* anything! Things will be *different,* he said, in *his* universe, where "almost all the matter within our sphere of influence will be serving the ends of intention rather than of static or structural support, which is mostly what it seems to be doing now. "

Matter was only *deadweight:* dull, inert stuff lying around passively and doing nothing at all. How lazy! How boring! How very thoughtless! Surely we can do better than *that!*

Indeed, why not just come out and *say* it: that we're going to march out there and subdue the whole universe, conquer the entire cosmos—all parts of it, without exception. That we're going to go everywhere, do everything, and learn all there is to know.

As a matter of fact, precisely that had already been stated, by John Barrow, the astronomer, and Frank Tipler, the mathematical physicist, in their book *The Anthropic Cosmological Principle,* which was published in 1986 by Oxford University Press. Now, Oxford University Press was the most reserved, traditional, conservative publisher in the history of the world, but it made no difference. At the end of the book Barrow and Tipler presented their vision of what it was going to be like way off in the distant future when mankind reached the "Omega Point," the point at which we finally would have . . . *Done It All.*

"At the instant the Omega Point is reached," they said, "life will have gained control of *all* matter and forces not only in a single universe, but in all universes whose existence is logically possible; life will have spread into *all* spatial regions in all universes which could logically exist, and will have stored an infinite amount of information, including *all* bits of knowledge which it is logically possible to know."

That was their simple, unpretentious program. The poor cryonicists, by contrast, who only wanted to store away a few human heads against the possibility of resurrecting them later on, why, they had an entirely reasonable agenda in comparison. Quite modest. *Humble,* even. They only wanted to raise the dead.

Fin-de-siècle hubristic mania, all right, and the progress of science was the key to it all. Just plain science. Just plain, old-fashioned, ordinary science would let us become the immortal spirits who will go out into the universe and liven up the place.

And why not? What else were science and technology good for if not to let us know what's knowable and do what's doable? No longer any place Mother Nature can hide.

So . . . on to immortality and the far-flung space culture. The two, indeed, were intimately related, for a world of immortals, or even a world of mortal Methuselahs, was going to get cluttered up pretty fast if there were no escape routes off into space. And so the first thing to get under way was the great space migration, the pilgrimage to the stars, the movement toward those dazzling "*all* spatial regions in all universes which could logically exist."

We'd have to start small, of course, with a few basic dance steps out into the solar system. Space colonies first, then on to the moon, thence to the asteroids, and finally off to the planets, and beyond.

If truth be told, we'd have to start by crossing even smaller distances.

1

Truax

*A*nd it's another fine day in the annals of manned rocketry. The countdown so far has gone exactly according to plan—"nominal," in the standard mission-control nomenclature—and it's now only about fifteen minutes behind schedule, which, in the rocket business, is close to perfection.

Little puffs of steam are venting off the side of the launch vehicle from where the propellant supply hose connects to the fuselage—nothing to worry about, though, just a normal part of the pressuring-up sequence. The steam puffs are blown away quickly by the wind that's been fanning down from the northwest all afternoon. Not much cooling in those breezes, for the temperature right now is hovering somewhere near ninety, but who cares? We're here, after all, not for our own comfort, but to see a rocket launch, to watch a human being ascend up into the heavens and, if he's lucky, return back to Earth again, all in one piece.

A safe distance away from the launch site, the audience—numbering some fifteen thousand, according to estimates—is listening to a Butte, Montana, high school band, resplendent in its silver and purple uniforms, play the national anthem and "Off We Go into the Wild Blue Yonder." Astronaut Jim Lovell is here to provide commentary, David Frost is doing live interviews, and Jules Bergman, ABC's crack science reporter, is covering the scene for "Wide World of Sports."

Those are the bare facts of the matter, but they don't begin to convey the emotional tone here at the launch site: the tension, the fright, the fluttery pit-of-the-stomach feeling like you get at the circus when they drop the net from under the high-wire act and the snare drums roll and the tightrope walker—one of the Flying Wallendas, maybe—steps off the platform and onto the wire with nothing on either side of him but yawning, empty space. It's that same sense of imminent danger, the realization that . . . *This man may die in the next few minutes!*

Only a few moments from launch now, but the crowd has not yet reached its ultimate pitch of high nervousness. That will come in a moment. At this instant all those present are waiting for the the grand arrival of the final necessary element, which is to say, the pilot, the human cargo that will shortly be rising up into the great blinding blue.

The man in question is now climbing into his jumpsuit. Not a pressure suit like those bloated Michelin-man outfits such as the astronauts wear during space walks, or even a "G-suit," the lace-up rubberized contraptions that fighter pilots wore back in World War II. No, the man's pulling on a piece of bright white double-knit, a costume that makes him look rather like Gunther Gebel-Williams, the lion tamer, only it's got the Stars and Bars running diagonally across the front. And where the usual NASA shoulder patch should go, there's a large number 1 instead. Shortly he'll explain the meaning of this.

He's standing in front of the TV cameras now, staring right at them, squinting into the sunlight, skin tanned, blond hair a little mussed up from the wind.

"I wear a red, white, and blue number 1 on my shoulder because I think I'm the best," he says. "In my business, you *have* to think you're the best, or you'll wind up dead."

Not the way an astronaut talks, certainly, but the man before us is not an astronaut. And he's not going into space either, although he is at this very moment (just like an astronaut) *walking out to his rocket.*

He won't be traveling very far, as rocket flights go: only a mile, more or less, across Snake River Canyon here in southern Idaho. Nevertheless, his low-altitude ballistic missile run will be just as much a manned suborbital flight as that of his only two predecessors in the sport, Alan Shepard and Gus Grissom. But whereas those fabled men had behind them the entire U.S. government, including the combined forces of Wernher von Braun and his rocket team at the Marshall Space Flight Center, plus the McDonnell-Douglas Corporation, the army, the navy, the Coast Guard, and NASA, today's hero will shortly be putting all his eggs, marbles, and family jewels into the hands of a single individual, a backyard rocket engineer by the name of Bob Truax.

*B*ob Truax must be a crazie—after all, he's got six kids. Just recently turned seventy, he looks like your old football coach: gray hair in a flattop crew cut, plaid shirt, khaki pants, chukka boots. A kindly man who speaks in a soft voice and in reasonable, well-constructed sentences, he has degrees in aeronautical and mechanical engineering, plus a master's degree in nuclear engineering, from, in his own words, "some jerkwater college out in the Midwest." (This turns out to be Iowa State University.) He's also done graduate work in biochemistry and microbiology at Stanford, this in pursuit of his other main interest, which is the eradication of aging and death. Truax, like lots of other extremely hubristic thinkers, regards getting older and dying as aberrant conditions, bodily malfunctions that could be corrected in the same way that any other human ailment is, through the application of good old-fashioned science.

But Truax's main passion has always been rocketry, and in fact he was the designer, inventor, and world's number one builder of the latest entry in high tech for the masses, *the personal spacecraft.* Truax wanted to do for rockets what Jobs and Wozniak did for computers: he wanted to make them into everyday items, machines that people could own, operate, and run by themselves, personally. Only instead of processing words or crunching numbers, you'd ride this machine up into the stratosphere, or farther. He's already built the prototype, the X-3 (the *X* is for "experimental"), also known as the Volksrocket, also known as Project Private Enterprise.

A home-grown assemblage of hope, hubris, and surplus parts, the X-3 is twenty-five feet long by two feet wide, and it will, when it's ready ("in about a year and a half"), carry the world's first private astronaut up into space. This will be "space" in the true, record-book sense, one hundred thousand meters up, or about fifty miles. After that comes the barnstorming era of space travel. And then the private space line.

One might of course regard the X-3 as the sheerest crackpottery, but then again Bob Truax was only pursuing the dream that NASA had been dreaming for the last thirty years or so: the fantasies of Men in Space, orbital colonies, a moon base, and Mars missions. It's the same story that Tsiolkovsky and Oberth and Goddard had been telling their contemporaries as far back as a hundred years ago, at the turn of the century, back in the last great wave of *fin-de-siècle* hubristic mania. There was supposed to be this vast human migration off the earth, into the solar system, out into the cosmos. And it even *began*, what with *Sputnik, Gemini*, the moon landings, and so forth, but *then* what happened? What happened was that NASA went the way of all other government bureaucracies, becoming a stultified, overgrown monster.

The space shuttle, a misconceived craft from the start in Truax's view, was supposed to be a cheap "space truck," hauling all manner of people and cargo up into orbit for twenty-five dollars a pound or some other such paltry figure. NASA said that a shuttle would go up every two weeks, and that eventually the whole operation would be turned over to private enterprise.

And *then* what happened? What happened was that the shuttle flew once every few months or so, putting up top-secret military payloads at the bargain basement price of eight thousand dollars a pound.

It soon became clear to advanced thinkers like Bob Truax that if We the People wanted to get up into orbit we'd have to do it on our own, privately. This didn't even have to be outlandishly expensive: Truax's asking price for putting the first private astronaut into space was a mere one hundred thousand dollars, roughly equal to the cost of a Fortune 500 weekend sailboat, light sporting aircraft, or modest hunting lodge (with Jacuzzi). And after the first private astronaut made it back safely, the price of the next flights could only come down, just like what happened with computers.

How symbolically fitting, therefore, that Truax lived in Silicon Valley, in Saratoga, California. To get to his house you took the De Anza Boulevard exit off the 280 freeway, exactly as if you were heading for Hewlett-Packard or Apple Computer only a mile or so away, not far from where Jobs and Wozniak put together the very first Apples, in Jobs's garage. Truax, admittedly, had a bigger garage than they did: his was a five-car, or more properly speaking, a five-rocket, garage.

Truax Engineering, Inc., is laid well back from the road, hidden behind a winding drive flanked by rows of Italian cypresses. The yard is a mixture of fallen tree leaves, cars, motorcycles, jet engine parts, rocket motors, and the like, with the X-3 itself up on concrete blocks. Farther back there's the Truax swimming pool, shaped like the state of California, and the keeshonds, which are raised by Truax's wife, Sally. The kids who still live at home—Scott and Dean—never cared much about the private rocketry business. "They think I'm nutty as a fruitcake," Truax said with a smile. Not that his kids were alone in this belief: "NASA doesn't talk to me. They probably think I'm nuts. Most people do."

Truax's fascination with rockets went back a long way. He first began making personal spaceships—small ones, anyway—as a boy in Alameda, California, where he and a friend fired off test models of their own design. This was in the late 1920s. The friend would carve beautiful specimens out of balsa wood, then Truax would fill their insides with gunpowder that he'd extracted from shotgun shells. He'd light the mixture with a fuse, run as fast as he could, and then await results. Almost always, the results were explosive.

So naturally they made bigger and better rockets, Truax concentrating on making an engine that wouldn't blow up on ignition. He experimented with all kinds of combustion chambers: paper straws, metal CO_2 cartridges, and just about anything else that would hold a charge of black powder. He even tried making his own solid propellants, alchemical mixtures of saltpeter, gunpowder, gum arabic, and other secret herbs and spices. "Some of them burned fairly decently," he remembered.

His greatest triumph, however, was the movie-film rocket. Truax got the idea when he found some movie film in the trash can behind a local theater. Celluloid was highly flammable and would, he decided, make a great rocket propellant, so he stuffed some of it into an empty tooth-powder can, applied the match, and stood back. The rocket was a complete success. "This one burst at a height of several feet and scattered strips of flaming celluloid all over my backyard."

To Bob Truax one of the major appeals of rocketry was always the sound of a good engine burn. "It's a very sexy sound," he said, "a very impressive sound. It's a high-pitched whoosh like a jet engine, but there's an unsteadiness to it, because when a supersonic jet comes out into a quiescent medium it creates a periodic turbulence. But it gives you an emotional response of some kind. Actually, I think a lot of the reason for being enthusiastic about rockets—for some it's a vocation, for some an avocation, and for some it's almost a religion—well, the noise has got to have something to do with it."

From the very beginning Truax had greater things in mind than mere noisemaking. He'd seen the future at an early age and wanted to be an essential part of it. These visions appeared in the afternoon newspaper, in the panels of a comic strip that he'd read almost from its first publication on January 7, 1929, "Buck Rogers in the 25th Century."

According to the story, Buck Rogers was a twenty-year-old flying cadet when World War I suddenly ended and he was forced into a job as a mine surveyor. One day while he was deep inside an abandoned mine shaft the roof caved in, releasing a gas that put him to sleep for the next five hundred years. When he finally woke up it was A.D. 2430 and people were floating through the air like butterflies, firing particle-beam disintegrator guns at each other, doing all sorts of incredible things. But all of it was explained scientifically. People could fly, for example, because they wore belts of "Inertron," a substance that resisted the pull of gravity: it had "reverse weight," the comic strip said. *It falls upward.*

These future people watched television, lived in vast cities of pagodalike skyscrapers covered by "metalloglass" canopies ("transparent, but with a strength greater than steel"), and traveled from place to place on conveyor belts or in airborne cars. If they wanted to, they could even put themselves into suspended animation by means of "neutro-metabolic tubes." They had robots, underwater cities, and—best of all—spaceships.

Spaceships! These fabulous rockets took people to the moon, Mars, and Jupiter, and to the enigmatic "Planet X," which was ruled over by a bikini-clad empress whose face the reader never got to see.

In 1930, at the age of thirteen, Bob Truax thought that all this was mighty interesting. Here were Buck Rogers and his decidedly nubile traveling companion, Wilma Deering, on their way to the moon. *It was amazing!* They were aboard a rocket moving along at eighty thousand miles an hour, and the two of them were just floating around weightless, like angels. And there was the memorable occasion when Wilma brought Buck a cup of coffee and the coffee floated right up out of the cup, *all in one piece!* But Buck Rogers wasn't at all daunted by this: he drank the coffee anyway, sipping it *right out of the air!*

Bob Truax got the idea that all this wonderful stuff was really possible, that it would actually happen someday, an impression that was only reinforced by the factual accounts of rocket experiments that he was reading in *Popular Mechanics* and in Sunday supplement articles. These told about the rocket test programs of Fritz von Opel and Max Valier in Germany, and about Robert Goddard, in Massachusetts, who had a scheme for actually flying to the moon, just like in Buck Rogers. All you needed, Goddard said, were staged rockets: one, two, three, and you were there.

All these dreams were still alive when, much later, Truax won an appointment to the U.S. Naval Academy. Now, by any standard Robert C. Truax was an extremely independent spirit, so one might offhand imagine that he'd be completely out of his element at a highly regimented military institution like Annapolis. But in fact he felt right at home there. Indeed, he *loved* the place.

Most of the plebes hated the first-year hazing, the strange disciplinary procedures and quaint rules dumped on them by upperclassmen, but not Bob Truax. He actually *enjoyed* it all. There were the rules for "straight corridors" and "straight turns," meaning that you had to walk down the exact center of any hallway and go this way or that only by stopping and doing a right-face or a left-face and then walking on. That was kind of spiffy. Then there were "square meals," a plebe punishment that involved geometric fork maneuvers at the dinner table: fork vertically up from the plate— stop—then horizontally to the mouth—stop—then reverse trajectory. It was a little crazy, of course, but still, it was fun.

He retaliated in the same spirit, going out of his way to play dumb jokes on upperclassmen. Truax's particular specialty was crawling under the dinner table and setting upperclassmen's shoelaces on fire.

Annapolis was a mixed blessing insofar as do-it-yourself rocketry was concerned. There was plenty of shop machinery available but not much time to use it, so Truax was forced to compress all of his hardware work into a frenzied half hour between the end of classes and the start of evening formation. About four-thirty each afternoon he'd get out of class and run at top speed over to the Steam Engineering Building, a half-mile away, where the machine shop was. Not only did he have evening formation to worry about, there was the added annoyance that the shop's electric power was regularly turned off at five o'clock. With only thirty minutes to use the lathe, drill press, and other equipment, Truax became a fast master of the TLAR ("That looks about right") school of rocket construction.

He put together a variety of thrust chambers and tested them out with different combinations of liquid and gas propellants. Because he couldn't do any testing on campus—the authorities feared he'd blow up the place—he had to go over to the naval experimental station across the river. Evidently, though, Bob Truax was making progress, because none of these new rocket engines ever exploded.

Truax wanted to fly, of course, and soon he was down at the New Orleans Naval Air Base taking flight training. He started off in the navy's "yellow perils," big radial-engine biplanes, some of which are still in use as crop dusters. He soloed after about nine hours and then went on to fly everything he could lay his hands on, even if it was for only an hour or so. Every once in a while a plane would have to be ferried from one airport to another, and Truax would always volunteer; it never mattered to him whether or not he'd flown that particular model before.

"One case I remember, when I was stationed in California they had an F4F they wanted to go to Jacksonville, and I wanted to get back East. Things were a lot more informal in those days, and so they gave me the engine log, the airplane log, a bunch of papers, they stacked me up with these things, and as I started out the door they asked, 'Oh, by the way, have you ever flown that thing before?' I said, 'Nope,' and they said, 'Well, why don't you take it around the field a couple times, practice a few landings?' So I took off, buzzed the field twice, and made my first landing in Phoenix."

By the time he left the armed forces in 1958 he'd flown in thirty-eight different types of aircraft. "I had less time in more planes than anybody else in the navy."

In 1939, shortly after graduating from the naval academy, Truax met Robert Goddard, who himself had been doing rocket experiments for the past ten years off in the obscure deserts of New Mexico. Goddard was a professor of physics at Clark College, in Massachusetts, where he'd done some of his first rocket testing, and had even conceived a plan for reaching the moon. He wrote this up in a book, *A Method of Reaching Extreme Altitudes*, which so horrified the *New York Times* that it saw fit to print an attack on the author, in an editorial that later earned for the paper a certain measure of "it-couldn't-be-done" distinction. Entitled "A Severe Strain on Credulity," the editorial claimed that rockets just wouldn't work in space, the reason being that they needed "something better than a vacuum against which to react." Goddard, they said, "only seems to lack the knowledge ladled out daily in high schools," as, for example, "the relation of action to reaction."

Actually, there *were* some things that Goddard apparently was ignorant of, at least according to Bob Truax. "Hell, Goddard was a physics professor," Truax said, "but still, all of his first rockets were *nose-drive rockets!* He figured that with the motor pulling from in front, the rocket would go straight up. *Ridiculous!* He should have known better than that! When a rocket moves, the thrust line moves with it, and in general it's only the *line of action* of the thrust that counts, not the point of application of the thrust along that line. That doesn't have anything to do with it."

Anyway, in the middle of World War II Goddard came to Annapolis, where he was made director of research on jet propulsion. His first assignment was to put rockets on the navy flying boat, the *Catalina*.

The *Catalina* was a beautiful ship with extraordinary range—it could fly about three thousand miles without refueling—but it had lousy takeoff characteristics and was totally unfit for aircraft carrier duty. But the navy brass wanted it on carriers, and so they decided to fit the ship with a JATO (jet-assisted takeoff) system. Since the navy wanted this in a hurry they put two separate groups to work on it: one would be headed up by Goddard, the other by Bob Truax.

Goddard planned on using gasoline and liquid oxygen propellants, but Truax, who had ample experience with these, knew right away that putting such things aboard aircraft was a risky proposition. Oxygen was liquid only at temperatures below −140°C and had to be stored in cryogenic tanks, which could cause all sorts of problems aloft. Truax, by contrast, planned to use his newly discovered nitric acid–and-aniline concoction. He'd been experimenting with these chemicals one day and had suddenly learned, much to his surprise, that they burst into flame spontaneously, upon contact, without benefit of an igniter.

That was a bonus: it let you turn a rocket engine on and off at will, like a switch. "I can remember Goddard watching one of my test runs, a look of absolute amazement on his face," Truax said. "Here we were, turning on, turning off, turning on, repeatedly, and there was Goddard having the devil of a time with his own nonspontaneous propellant."

Truax's team was the first to get the *Catalina* into the air with a jet assist, although Goddard had made several tries at it. The plane was fitted out with Goddard's rocket units and made six runs down the Severn River, but salt spray and vibration shorted out the igniters every time. On the seventh try, a liquid oxygen line came loose, causing a fire that damaged the aft end of the plane. After it was repaired a few months later, Truax put his own rockets aboard, attaching them to the wing struts both for safety reasons and to make them reusable. Once in the air, the pilot could jettison the rockets, which could then be retrieved from the water.

With Truax's JATO units going full blast, the *Catalina* got off the water on the first run. His rockets were so powerful, in fact, that they launched into the air planes that were too heavy to maintain level flight with only the thrust of their own engines.

After the war, Truax went on to dominate the military rocket program. He developed the Thor rocket for the navy, came up with the idea for a submarine-launched missile, the Polaris, and invented the type of engine that was used on the Bell X-1 rocket plane, the one that Chuck Yeager flew across the sound barrier. Then, even before *Sputnik* was launched in 1957, Truax was head of the Air Force Space Program, a top-secret and short-lived effort to beat the Russians into orbit. By the time he quit government in the mid-sixties, Bob Truax had become one of the world's foremost experts in rocketry.

The idea for a canyon jump came to Evel Knievel in a bar called Moose's Place in Kalispell, Montana. He was drinking a mixture of beer and tomato juice ("They call that a Montana Mule up there," he said), and staring at a picture of the Grand Canyon that hung on the wall. The more he drank, the smaller the canyon looked, and before long he'd decided to jump it.

That was in 1966 or so. By this time Knievel already had scores of motorcycle jumps to his credit, over lined-up rows of cars, buses, Mack trucks, boxes of rattlesnakes, open cages of mountain lions, and so on, and success at this kind of thing could easily give a person some big ideas. Not that these jumps were always successful. Knievel had done three hundred of them by this point, and had crashed eleven times, getting some fifty-odd broken bones in the process. Nevertheless, a trip over the Grand Canyon seemed like the next logical step.

After a while he even got to imagining how it would go. He'd picture himself on a jet-assisted Harley Davidson streaking down a long approach road lined on both sides by countless millions of people. They'd be smiling, waving, cheering him on: "Go, Evel! Go, Evel! Go, Evel!" He could just barely hear these screaming voices over the din of his engine, but he'd have this big, wide grin on his face, and he'd nod his head in acknowledgment and smile and wave back as he went by. Then he'd come to the takeoff ramp, the long, inclined platform that would lift him up to just the right angle to go flying up, out, and over the gorge. He'd cut in his rocket engines, and *Zooooom* . . .

It sounded like such a great stunt that he finally wrote a letter to Stewart Udall, Secretary of the Interior of the United States of America, asking for official permission to cross the Grand Canyon by air, on a motorcycle. Udall didn't say yes, but he didn't say no, either, at least not right away, so for the next two years Evel Knievel went around talking about the flight and making arrangements.

Or at least he tried to. He had to get a building permit for the takeoff ramp from the Navajo Indians, who had authority in the matter, but they said no. And then Stewart Udall finally wrote back and said no, too. He had decided after long reflection, he said, that the public lands of our great country really shouldn't be used for this kind of thing.

That was the end of the Grand Canyon jump, but there were plenty of other broad and deep ravines around in the desert Southwest, and after a while Knievel located a privately owned cattle ranch in Idaho on the edge of Snake River Canyon. He leased it from the Ted Qualls family, the owners, and then got serious about jet propulsion.

While the jump was still in the daydream stage, Knievel bought a couple of surplus rocket engines and attached them to a motorcycle. He put this rocket-powered bike on display when he did his "normal" jumps, billing it as his canyon-jumping machine. One time an aeronautical engineer named Doug Malewicki happened by and saw it and was shocked. He knew right away the thing would never work. For one thing, a canyon jump would require a takeoff speed on the order of four hundred miles per hour, and at that rate the rider would be blown off like a speck of dust. And for another, a flying motorcycle would be dynamically unstable: it would start tumbling end over end almost as soon as it left the ground. The only way to do a jump like that, Malewicki thought, was with a rocket. You needed aerodynamics, you needed wings.

Malewicki wrote up these observations in a letter and mailed it to Evel Knievel. After a while Knievel asked Malewicki to build him a canyon-crossing rocket ship.

Malewicki knew aerodynamics, so he thought he could design the body of the ship, but he knew nothing about rocket engines. He showed up one day at the offices of Aerojet-General, commercial rocket manufacturers, in Sacramento. It seemed as if they made just about everything that went up into space: the Aerobee high-altitude research rocket, the Titan III-C solid rocket boosters, and so on and so forth. Malewicki went in and calmly explained that he needed a rocket engine for a short motorcycle trip across a canyon.

Aerojet agreed to look into it, and in fact even quoted him a price, which was in the hundreds of thousands of dollars. That was more than even Evel Knievel could afford at the time, but then someone at Aerojet happened to mention Truax. "He's been fooling around in his garage with steam rockets," this fellow said. "Maybe he could put together something in your price range."

Truax had come to Aerojet after he left the military. The company put him in charge of its Advanced Developments Division, gave him a millon-dollar-a-year budget, and told him he could spend it however he liked. He spent the money figuring out how to make large rockets cheaply—"Big Dumb Boosters," as he called them. That effort was fine so far as office work went, but Truax missed the experience of actually building rocket engines with his own hands—he needed to hear that fevered roaring, that sexy periodic turbulence—so he was always looking to do an extra bit of home-brew rocketry. As in the case of the rocket-powered racing car, for example.

In the mid-1960s, Walt Arfons, the race-car driver, was looking for something new in racing cars. Truax suggested a steam rocket.

A steam rocket was a simple device, nothing but a pressure cooker with a nozzle at one end. You filled it with water, super-heated it under pressure, and then pulled the plug out at the end: the steam rushed out and the craft moved off in the other direction. It was an old idea, and in fact such a rocket had been patented in England as far back as 1824. One of the great virtues of the steam rocket was safety: since it used no flammable propellants, it was impossible for it to go up in flames.

Walt Arfons took Truax's steam rocket, put it in a car, and gave a press demonstration at the Akron, Ohio, airport, near where he lived. Truax had made the engine throttleable: the power was controlled by a foot pedal, just like in any normal car. He'd told Arfons to work up to speed gradually, but on its first public test run the driver got carried away, floored the pedal, and then held on for dear life as the car left the ground and went through the typical crash-and-roll sequence. In the process, a valve broke open and hot water shot out in a beautiful spiral arc, but nobody got hurt in the least.

Anyway, when Doug Malewicki stopped in to see Truax about an engine for a canyon-jumping rocket, Truax realized that a steam rocket would be just what the doctor ordered, so he built Malewicki a copy of the race-car engine.

Malewicki, meanwhile, had hired out the bodywork to a hot-rod fabricator who ran a shop out on the West Coast. One time Truax went out there to see the thing, which they were calling the X-1.

"This X-1 was a beautiful-looking machine, but it was structurally unsound. I took the vertical fin in my fingers and moved it across a two-inch arc, just like that. I asked the builder what it was hooked to underneath, and he said, 'Nothing, it's just hooked to the skin, to the outer skin of the rocket.' Well, I knew that would come off in a second."

Eventually, Knievel scrapped the X-1 and put Bob Truax in charge of the whole flight. Truax designed and built not only the engine, but an entirely new rocket, the X-2.

The X-2, like most of Truax's homebuilt designs, would be a bunch of surplus parts flying in close formation. The hot-water tank was an oxygen bottle from a B-29 bomber; the nose cone was a tip tank from a Grumman Goose flying boat; the parachute release gyroscopes came from a Nike-Ajax missile. Truax invented the rest of the craft himself, including the parachute deployment system, a complex Rube Goldberg mechanism that involved a gun that fired a slug that tore the lid off of the parachute canister. The drogue chute was attached to the canister lid, and when the lid came off, the chute would be pulled out by the slipstream. As complicated as it was, the system worked, every time.

Eventually, Truax made two copies of the X-2—one for a pre-flight test, the other for the real thing.

*T*he locale has changed from the Grand Canyon to the Snake River, and the motorcycle has been replaced by an honest-to-God rocket ship, but the rest of the nightmare has gone according to plan. Now, at three in the afternoon of September 8, 1974, Evel Knievel is (just like an astronaut) *walking out to his rocket.*

Dubbed the "Sky-Cycle X-2," the vehicle in front of him doesn't resemble a motorcycle in the least, and in fact consists of little more than nose cone, open cockpit, fuel tank, and wings, in that order. The wings, such as they are—mere horizontal and vertical stabilizers at the aft end—give the craft a slight resemblance to the X-15 rocket plane. As for control surfaces, there's a pair of small fins up front ("flippers," Truax calls them) that the occupant can manipulate by means of floor pedals. A galaxy of white stars has been painted across the top of the craft, and on the side the name EVEL KNIEVEL appears in large, gold block lettering.

The X-2 rests at the bottom of a 108-foot-long steel ramp, which is pointed at fifty-six degrees up from the horizontal. When the rocket engine fires, the X-2 will go from zero to four hundred miles per hour in one fell swoop of acceleration a mere four seconds long. The apex of the flight will be about three thousand feet above ground level, roughly the height of three Empire State Buildings, and a vantage point from which, if he had the least momentary interest in sightseeing (and if he hadn't blacked out from the G-forces), Evel Knievel could peer over the side of his craft and behold the fine desert landscapes of three states—Idaho, Utah, and Nevada—spread out far below.

Right there, at the top point, Knievel is supposed to let go of a switch, thereby activating the Rube Goldberg parachute mechanism that would stop the Sky-Cycle X-2 dead in its tracks and then lower it to the surface at fifteen miles per hour. Five Gs positive at launch, followed by three Gs negative when the chute snaps out— Eyeballs *in!*, Eyeballs *out!*—all of this in about sixty seconds.

And then there's the wee matter of landing to engage your attention. During the course of his career travels aloft Evel Knievel had gotten quite accustomed to impacting all sorts of immobile objects: dump trucks, cars, buses, cement walls . . . whereas this time it will only be good old Mother Earth, but *still!* What if he goes in the water? *He can't swim!* What if he hits the side of the cliff? *He could bump and crash his way down for six hundred feet!*

No, this is not going to be any trivial impact, and the man himself is already beginning to feel the strain. "Right now I don't think I've got better than a fifty-fifty chance of making it," he says at a prelaunch press conference. "It's an awful feeling. I can't sleep nights. I toss and turn and all I can see is that big ugly hole in the ground grinning up at me like a death's head."

And who can blame him? "If the heater doesn't blow up and scald me to death on the launch ramp, if the countdown goes right, if the Sky-Cycle goes straight up and not backward, if it actually reaches two thousand feet, if the chute works, if I don't hit the wall at four hundred miles per hour, and if I can get out of it when it lands—I win. If it doesn't work, I'll spit the canyon in the eye just before I hit.

"Then again, I've got five backup systems," he added. "The fifth one is called the Lord's Prayer."

The entire system has undergone only an extremely limited amount of flight testing. Lacking a wind tunnel, Truax ran his stability tests by attaching a scale model of the X-2 to the front fender of his Chevy El Camino pickup and driving it up and down Route 108.

Then there were flight tests of the two full-size vehicles, the scrapped X-1 and the duplicate-copy X-2. The first flight, made in November of 1973, was done mainly for publicity value, back when Knievel was trying to sell the film rights. He was hoping the rocket would land in the water, because that would make the whole enterprise look deadly and thereby get him bigger fees. From an engineering standpoint, Truax wanted to make sure that the rocket would stay on the launch ramp tracks and not go over the side halfway up. He used the X-1 for this purpose, the one made by the hot-rod shop. Because the craft had no autopilot or guidance system, Truax cut the engine burn short, to prevent the rocket from circling back around toward the test stand and wiping out himself and everyone else. So, like Knievel, Truax was hoping that the X-1 would wind up in the canyon.

Which it did. When launch time came, steam sprayed out of the exhaust nozzle and the craft streaked up the ramp like a bullet. It rose into the air a few hundred feet, arced over, began a series of uncontrolled rolls and tumbles, then finally straightened itself out somehow and went into the water at midriver, making a big splash. From everyone's viewpoint, the test went perfectly: the craft plopped in at the exact center of the river, precisely on target.

The other test was with the duplicate Sky-Cycle X-2. This was a dress rehearsal for the actual flight, now only two weeks away, and this time the rocket was supposed to make it all the way across to the other side. The duplicate X-2 had an autopilot, parachute system, and all the trimmings, even including, for an added touch of realism, a mannequin ("Fred Galahad") strapped into the pilot seat.

So they went through the checklists and the countdown, the engine spurted out its blast of steam, and the craft shot up the launch ramp. But before it even reached the *end* of the ramp the drogue chute popped out! The rocket stopped dead in its tracks, as if it had suddenly flown through a wall of cotton, then nose-dived straight down and was lowered gently to the bottom of the canyon. The X-2 had a shock absorber at the front, and when it went into the mud nose-first it stayed there, stuck like a dart.

That was a rather bad moment for all concerned. Except, ironically, for Evel Knievel himself, who would be in the very next shot. Truax wanted to postpone the flight until he could locate the parachute problem and make corrections, but Knievel just wasn't listening. For one thing, these two near-disasters would make the actual flight look even more risky than it was, and that would bring in the crowds. For another, the blunt fact of the matter was that he would have survived: Fred Galahad was hardly scratched.

But a third factor was even more decisive. "Hell, I'll lose two million if I postpone it one day," Knievel told Truax. "I'm gonna go."

To which Truax said, "Okay, it's your life and your money."

Truax and his crew now worked overtime trying to discover why the chute had popped out. They ran a bunch of static tests with the second X-2, the "real" one, the one that Evel Knievel would actually ride across the canyon. With the rocket held captive by a hold-down mechanism, they fired the engine again and again. Sometimes the chute came out prematurely, sometimes it didn't. After a while it became clear what the problem was.

There were two umbilical cables running to the rocket, one for sending commands and the other for taking instrument readouts. It turned out that if the command cable was attached and the instrument cable wasn't, then the chute stayed where it was, but when *both* cables were attached the parachute erroneously deployed. Clearly there was a fault in the wiring: something was transmitting a bad signal back through the instrument circuits, mistakenly triggering the chute. The problem was locating the right circuit, but there were hundreds of feet of wiring in the rocket and no good way to discover exactly where the fault was, especially not in the last few days remaining before launch.

"So I told them to rip out all the electrical stuff," Truax said. "This was heartrending, really, because there was an instrumentation package in there with a flight recorder—a 'black box'—so we could postmortem the flight if something went wrong. There was the parachute-release gyro, there was the autopilot to keep the vehicle from rolling in flight, and all of it had been wrapped up beautifully. The wires were encased in conduits and so on, it was an absolutely beautiful job. But something in there was giving a bad signal, so I told them, *Tear it all out.*"

With the automatic parachute-release mechanism gone, Truax had to jury-rig one that Knievel could control manually, from the cockpit. He settled on a dead man's lever: as long as the pilot kept his hand on it, nothing would happen, but as soon as he let go, the chute would deploy. It was crucial that Knievel let go at precisely the right moment: too early and he might not make it across, too late and he might fly halfway to Canada.

"You watch the horizon out there in front of you," Truax told Knievel. "When you see more earth than sky, then you let go of the lever."

Since Knievel had a student pilot's license, Truax thought he would be able to tell earth from sky. But even so, Truax gave him a backup system (just like in NASA), a stopwatch with a red sector painted on its face.

"When the needle gets into the red sector," Truax told him, "let go of the stick. That's if you haven't already done this earth-sky business."

T he previous days and weeks have given everyone plenty of time to wonder about the larger meaning—if any—of the event we're now about to behold. The facts of the matter are clear enough: Evel Knievel is going to ride this candlestick up into the sky to where—if the rocket engine fires for just a second or two longer than it's supposed to, and it's supposed to fire for precisely four seconds, no more and no less—he might actually *disappear from view!*

During interviews and at press conferences, reporters continually asked Knievel, *Why are you doing this?* and he always gave them the same answer. "I like to live with a lump in my throat and a knot in my stomach," he said. "A man is meant to live, not just survive."

Bob Truax, on the other hand, simply wanted to build more launch vehicles. "I can't get enough of it," he said. "I just like to go out and play with rockets."

And as for the on-site crowd and the home viewing audience, why were they watching, other than to find out whether this latest example of flagrant hubris would be punished, as it always had been from time immemorial? The more classically educated among them might remember the tale of Icarus, the boy who flew out over the Aegean Sea on wings made of feathers and wax. He flew up toward the sun, toward where it was warm, up to where it was in fact *too hot*, so that his wings melted and he fell into the water and drowned. That was what you got for hubris.

They might think of Prometheus, whose very name meant *forethought;* they might remember how he stole fire away from the gods on Mount Olympus, brought it down to mankind, and then, in return for this effrontery, was chained to a rock, where the vultures alighted to feed on his liver. Every night his liver regenerated, and the birds came back again the next day. This went on endlessly, for centuries, until finally Hercules killed the vultures and set Prometheus free. Those too were the wages of hubris, from which the human race was supposed to have learned one important lesson, to wit: *Hubris does not pay!*

But some of us have never learned that lesson, as becomes clear when Evel Knievel's Lear jet (rented specially for the occasion) comes sweeping down the canyon from right to left. This is the signal that the hour is here, that time zero is upon us.

Knievel, dressed in his white jumpsuit with the Stars and Bars, arrives on the scene and the crowd gives a loud whoop. He acknowledges this with a wave of his $22,000 gold-and-diamond-tipped walking stick, then walks over to the speaker's platform.

Inside the launchpad control center, also known as the "Sky-Cycle X-2 Super Van," Bob Truax realizes that the propane furnace that heats up the rocket's water has just about given up the ghost. It got the water up to temperature once—to its superheated 475°F—but probably can't be made to perform that miracle again. The water will hold for about thirty minutes, but after that there will be no flight.

Knievel, meanwhile, has made it to the speaker's platform out by the launch ramp. The platform is a wooden stand that, from the crowd's point of view, looks as if it's jutting out over the canyon rim, although in fact it's standing right at the edge, dwarfed by the void beyond it. Up on the stage, television personality David Frost, a Catholic priest, and assorted officials now enact the final prelaunch ceremonies.

Evel Knievel addresses the crowd. The loudspeakers make him sound like he's calling in from somewhere around Jupiter, but it's questionable whether what he's saying would be comprehensible even if you could hear it. When he's finished he steps back from the mike and the priest advances to murmur a few words of bene-diction—something about "a man with a dangerous dream," about "happy landings whether on earth or in heaven," and so forth—in a steady, garbled drone. He pronounces a few more blessed sen-tences—*garble, garble*—and then a few more. This goes on for a while. Finally Truax sends a runner up to the platform.

"Christ's sake!" Truax says to the runner. "Tell them to shut up.

The water's cooling off and he's going to land in the river if this goes on any longer."

Knievel gets the message and now prepares to initiate his cockpit ingress sequence. Rather than simply climb down from the speaker's platform, walk to the end of the launch ramp, and climb up a ladder to the rocket—which to an ordinary person is the only conceivable way of making the trip—Knievel has chosen to add an extra measure of dramatic tension to the proceedings and so now an enormous construction crane that's been standing by the whole time guns its engines and starts moving its long, blue boom over toward the platform. It lowers a bosun's chair—a mobile astronaut's seat—and Knievel hops onto it, then swings out over the crowd.

And now the crowd, which has been raucous and milling and muttering and swilling beer all afternoon, is suddenly struck dumb, every last person—the VIPs, the press photographers, the bikers, the clean-cut families from Wisconsin, the lustrous, T-shirted teenage girls—all of them slapped into submission by the harrowing realization that . . . *This man may die in the next few minutes!*

Knievel, gyrating around like a Foucault pendulum, reels out over the crowd on the way to his rocket and gives a brave thumbs-up—*The courage of the man!* ("If it doesn't work, I'll spit the canyon in the eye just before I hit," he'd said!)—and hundreds of thumbs-ups are silently returned from below.

At 3:20 P.M., Evel Knievel lowers himself into the X-2 and fastens his lap belt, shoulder straps, and parachute harness. He grabs on to the parachute-release stick and pronounces himself ready to go. Truax goes through a checklist with his assistants, then leaves for the launch van.

At the control console Truax stays in touch with his crew by walkie-talkie. The countdown is broadcast to the throngs outside, Truax yells, "Get clear!" into his walkie-talkie, and Evel Knievel, invisible except for his red crash helmet, is alone with his thoughts, the knot in his stomach, and the lump in his throat.

*W*HOOOOSHHHHH! . . . A burst of white steam and fulmination and the X-2 whips up the track. But even before it reaches the end, *the drogue chute pops out!* . . .

(*Damn, he let go of the stick!* Truax says to himself)

. . . and then the main parachute comes streaking out after it. But the rocket keeps on rising until, all at once, the engine shuts down and the X-2 pitches over and begins a slow roll to the right. It's a stunning roll, absolutely air-show quality, one that any aerobatic pilot would be proud to have pulled off—only the X-2 is doing it *all by itself,* because its autopilot had been ripped out.

The next moment the rocket's pointing straight at the ground, followed by twin spirals of red smoke. Truax had attached pyrotechnic smoke canisters to the ship so that he could keep track of the thing if it got too high and went out of sight. Even now, even though it had been held back by the parachute, the craft is only a tiny red-and-white blot against the blue sky.

Looking through his binoculars, Truax sees Knievel struggling, as if he's trying to unharness himself and jump from the rocket. They had always told him *not to do this:* "Don't get out of your harness until you come to a stop. Otherwise you'll wind up in the forward section, a mass of hamburger."

"Believe me, I'm not going to panic," he'd said at the time, but Knievel, who's wearing a chest-pack parachute, can hardly swim, and one of his worst fears had been landing in the water and being dragged down under. In the cockpit, Evel Knievel now sees the world go topsy-turvy, then he realizes that *he's made it,* that *he's over the other side,* but in the next instant he sees that he's *drifting back,* and then that he's *over the river again*—which now starts rushing up toward him.

Knievel raises his arms, as if to lift himself up and out of the cockpit, but Ron Chase, the ground communicator, radios to him, "Stay with the bird! Stay with the bird! It looks like you're going to go into the canyon, but you've got a good chute!"

But Knievel's still flailing his arms around wildly, and for the sake of getting his hands away from the seat-belt release, Ron Chase tells him: "Put your visor up! Raise your visor!" (The visor is tinted, and underwater it would probably block all vision.) Knievel stops whatever he's doing and tries to raise the visor on his helmet, but . . . *it's locked!* He rips it off, nicking his nose in the process.

The crowd, standing well back from the canyon rim, sees that despite everything, despite the drogue chute and the main chute, the rocket has actually made it across to the other side of the canyon. *But the wind!* The wind's pushing it *back . . . back over the water!*

"He's going in the river!" the crowd screams. "*Oooooooooooooh!*"

"*I* plan to arrange, if I ever get enough money, to have my head frozen," Bob Truax said. Indeed, going into space was only one of Truax's three major life projects.

"There are only two or three things I want to do in life. The first thing is, I want to cut the cost of space transportation down to the point where it's affordable, where we can really *do* things out there, where the cost of just getting *out* there won't be a barrier. Then I want to eliminate aging. And then I want to eliminate war."

Without a doubt, ending war was the hardest of Bob Truax's three projects; in any case, it was the one that he spent the least time working on. Unlike war, aging was a *scientific* problem and was therefore in principle solvable.

"I think perhaps that had I gotten interested in aging rather than rockets as a youngster," he said, "I might have solved the problem by now."

What he could never figure out was why it took scientists so long to recognize that aging was a problem they ought to be working on.

"It almost seems that the most important problem that the human race has ever had is the one that they're the slowest in attacking in a scientific, rational fashion. I don't understand it. But yet I've got to admit that I didn't think of attacking the problem myself till maybe twenty years ago. I don't even know what started me thinking about it. Maybe I began to worry that my time was coming. Actually one of the things that got to bugging me was the amount of information coming along that I had no time to master. I just couldn't get around to reading all the articles that I wanted to read. It's too bad, but the way things are now, a guy who wants to make a contribution has to go to school until he's forty years old before he gets to the point where he knows enough to make any kind of advance, and then twenty seconds later he keels over dead with a heart attack. It's a hell of a note, really."

As it was, Truax was in his late fifties before he got interested enough in the subject to actually *do* something about it. That's when he started his postgraduate work at Stanford, to try to figure out why it was that people aged, how you could prevent it, and how you could reverse the process. If there's anything that Bob Truax was sure of, it was that aging had to be caused by *something*.

"I believe in the law of cause and effect, so there's got to be a cause for every effect. If we discovered the cause of aging, then presumably, or possibly, we could come up with the cure."

For a while Truax thought that aging might be caused by the presence of deuterium—heavy hydrogen, an isotope of ordinary hydrogen—in drinking water.

"Hydrogen comes in two forms, normal hydrogen and deuterium, which is actually twice as heavy. It's present in normal water in about seven-tenths of 1 percent, or something on that order, a very small concentration. Nonetheless, it could be significant *if* it accumulated over time. Chemically, deuterium is almost identical to normal hydrogen, and when you link it with oxygen the difference in weight is not great. But the dynamics of the chemical reactions are quite different, and the strength of the hydrogen bond with deuterium is different."

Truax's theory was that one or more of the body's organs—the hypothalamus, for example, a part of the brain that governs metabolism—was being interfered with by the buildup of deuterium. Too much deuterium might slow down the cell-repair process, leading to aging and finally death.

He did some research into this and found that others had already studied the effects of deuterium-rich water—also called "heavy water"—on the growth of bacteria, plants, and animals.

"Some bacteria grew to *enormous* size," Truax said, "a hundred times their normal size. Then they tried plants, and they generally found that plants grew *less* well, that they were stunted when grown in high concentrations of deuterium. Then they got more of it and tried to deuterate animals. They grew food that also had high concentrations of deuterium in it, which was hard to do because of its effect on plants, but they finally got an algae that they could make into a cake, and they fed it to animals. There were all kinds of effects, but it turned out that there was no effect on the rate of aging."

At first, Truax thought that this invalidated his whole hypothesis that deuterium caused aging. But then he saw that there was a possible escape clause.

"It turns out that nearly all body chemistry is enzyme-mediated. That is to say, there is always a catalyst present that makes the reaction go, and controls the rate at which it goes. But there's a very fundamental law that states once you've saturated the catalyst, then there's no further change in the rate of reaction. Now no one had ever performed the *reverse* experiment, which is to raise some kind of a varmint on deuterium-*free* water. That would be the clincher."

Truax thought of doing the experiment himself, and for a while he tried to find some deuterium-free water, but found that it was hard to get. "I've heard something to the effect that there is some available in France," he said, but he never followed up on this.

He did, however, follow up on the progress of cryonics, and even joined the Bay Area Cryonics Society. Having yourself frozen was, to Bob Truax, a reasonably attractive interim measure, a way of keeping yourself alive (as it were) until such time as the causes of aging and death had been discovered and reversed. Of course there were no guarantees about the success of cryonics. "I asked them what do you do in the event of atomic war," he said, "but they made no provision for that. They can take care of temporary power failures, but not anything as serious as nuclear war. But I figure that's almost guaranteed to happen in the next hundred years."

The cryonicists, for their part, were well versed in the ways of space travel—"suspended animation" or "frozen sleep" being the fixture it was on interstellar journeys, at least in science fiction. In any case, they wanted to know about Truax's recent advances in rocketry, and so one time they invited him to give a lecture at one of the Bay Area Cryonics Society's meetings in Berkeley. Truax came and told them about the X-3.

The X-3 project began when Evel Knievel came up out of the canyon, unharmed except for the cut he got on his nose when he jammed his visor up. Knievel claimed then and forever afterward that he never let go of the stick, and Truax, after inspecting the rocket, agreed, deciding that the parachute mechanism had failed on its own. Anyway, when Knievel climbed up out of the canyon and saw Truax standing there, the first thing he said was, "Well, Bob, that's going to be one hell of a hard act to follow. What else you got up your sleeve?"

Truax had already given the matter some thought. He was impressed by the way Knievel's daredevil acts generated truly massive cash flows. Others were similarly impressed, and soon enough Truax was inundated with all sorts of suggestions for follow-up ventures. A group of Japanese businessmen, for example, wanted to know if Evel Knievel could rocket over Mount Fuji. They even flew Truax over there to assess the matter.

"Technically, it could be done," Truax told them, "but not economically or efficiently." Knievel, though, was always ready: "If Truax says go, I go."

But Evel Knievel would never make an assault on Mount Fuji. Truax had even better things in mind, so when Knievel asked him what they'd do next, Truax's answer was, "Well, if you can scare up about a million dollars, I think I can make you the world's first private astronaut."

It was an altogether reasonable proposition to Knievel, who'd known a few astronauts in his day and already had a hankering to join the club. He gave Truax a small "research grant" of about three thousand dollars to see what he could find out about costs and so forth, but not long afterward Knievel dropped out of the project entirely. He'd gotten into an unfortunate and expensive fray with an associate and no longer had a million dollars for this or any other purpose. But Truax went ahead on his own, for the project appealed to him on several levels. For one thing, launching the world's first private astronaut into space looked to be the ultimate in amateur rocketry. A single individual—he, Bob Truax— would challenge the mighty gods of outer space, and would prevail, with no government help whatsoever, no NASA, no military, no nothing, just his own ingenuity and spare parts. It was a great idea, probably the single best thought he'd ever come up with.

And there could be some money in it too, what with sales of

TV and film rights, the book, the magazine articles, all sorts of subsidiary rights, residuals, and God only knew what else. He once asked ICM—International Creative Management, publicists, author's agents, deal maker to the stars—to estimate how much could be brought in from a private astronaut shot. Not much, they said: only $10 million or $20 million.

Then, too, there was another angle, a more serious one. If it was successful, the private astronaut shot could be the *Kitty Hawk* of space travel. It would demonstrate that going into space didn't *have* to be an abnormally expensive undertaking. If a single individual acting on his own could do it, then why not others? Why not private astronaut *corporations,* private *space lines?* The stunt would be a way of getting the Great Space Migration rolling, of getting on with the whole Buck Rogers scenario. Besides, if he could do the shot the way he wanted to—launching it from the water, recovering the vehicle, and then using it again—then he'd vindicate the idea that he'd for a long time regarded as his own personal baby, the *Sea Dragon.*

The *Sea Dragon* was a launch vehicle of stupendous proportions that Truax had designed back when he was director of advanced development at Aerojet General. The best perk of that high office was the $1 million budget that he could spend any way he wanted to. Truax used it to test his pet theory that the *cost* of a rocket had nothing to do with how *big* the rocket was. You could make a given rocket just as big as you pleased and it would cost about the same as one that was about half the size, or even smaller.

This went against conventional wisdom and common sense, but at Aerojet Truax collected enough facts and figures to prove its truth beyond a doubt. Indeed, he'd been assembling the necessary data from the time he was still in the navy, where he'd had access to all sorts of cost information.

Take *Agena* versus *Thor,* for example. These two rockets were identical in every way: each of them had one engine, one set of propellant tanks, and so forth; the only significant difference between them was size. The *Thor* was far bigger than the *Agena,* but the surprise was that the *bigger* rocket had cost *less* to develop.

"I was shocked to discover the *Agena* cost more than the *Thor,*" Truax said later. "The *Thor* was between five and ten times as big!

I said to myself, We've been tilting at windmills all this time! If all rockets cost the same to make, why try to improve the payload-to-weight ratio? If you want more payload, make the rocket bigger."

The same anomaly cropped up again in the case of the two-stage *Titan I* launch vehicle: the upper stage was *smaller*, a miniature version of the lower stage, yet the smaller one had cost *more* to make.

It seemed irrational, but all of it made sense once you went through the costs item by item. Engineering costs, for example, were the same no matter what the size of the rocket. "You do the same engineering for the two vehicles, only for the bigger rocket you put ten to the sixth after a given quantity rather than ten to the third or whatever," Truax said.

The same was true for lab tests. "The cost of lab tests is a function of the size of your testing machine and the size of the sample you run tests on, not the size of the product."

Ditto for documentation: paperwork, spec sheets, manuals, and so forth. The cost here was a function of the *number* of parts and not the *size* of the parts. "There are absolutely no more documents associated with a big thing than a small thing, as long as you're talking about the same article."

By this time Truax had accounted for a healthy chunk of the total cost of a given launch vehicle. About the only thing that *did* vary directly with a rocket's size was the cost of the raw materials that went into making it, but raw materials constituted only *2 percent* of the total cost of a rocket. "Two percent is almost insignificant!" he said. "And even with raw materials, if you buy a ton of it you get it at a lower unit price than if you buy a pound. And this is especially true of rocket propellants."

So if all this was true, if engineering, lab tests, documentation, and so forth didn't determine a launch vehicle's price tag, then *what did?* Essentially, three things: parts count, design margins, and innovation. Other things being equal, the more parts a machine had, the more it was going to cost. The more you wanted it to approach perfection, the more expensive it would end up being. And finally, the newer and more pioneering the design, the more you'd end up paying for it.

"We came up with a set of ground rules for designing a launch vehicle," Truax said. "Make it big, make it simple, make it reusable. Don't push the state of the art, and don't make it any more reliable than it has to be. And *never* mix people and cargo, because the reliability requirements are worlds apart. For people you can have a very small vehicle on which you lavish all your attention; everything else is cargo, and for this all you need is a Big Dumb Booster."

Bob Truax's *Sea Dragon* was a Big Dumb Booster, an absolutely titanic launch vehicle, one that would weigh forty million pounds at lift-off. The *Saturn V* rocket, by contrast—the one used for the Apollo moon flights, and at that time the biggest rocket ever launched—weighed in at a paltry six million pounds. The *Sea Dragon* would be the *Spruce Goose* of space travel, so big that it would have to be built in a shipyard, and both launched and recovered from the water. After being hauled out from the ocean, the *Sea Dragon* would be refurbished and then sent back up into space. It would be a true "space truck," as opposed to NASA's space shuttle, which was then on the drawing boards.

In Truax's view, the space shuttle philosophy had everything bass-ackwards. Number one, since most of the time the shuttle was only going to be launching satellites (which could be done far more cheaply and efficiently with unmanned vehicles), there was no need to put people aboard. People were unnecessary, and they only made for tougher design margins, and therefore greater expense, and even *then* you had no guarantee that the crew would always return safely, which they sometimes didn't.

And number two, although it was supposed to be reusable, the shuttle was designed to land at an *airport,* as if it were an actual *aircraft,* which to Bob Truax was a laugh. "It makes about as much sense as requiring an airplane to be able to land at railroad stations," he said. "It flies like a brick and has a dead-stick landing, the most difficult of all landings. It's an unparalleled money sponge. But what are you going to do? *Mitzi and Ritzi want to go to the stars. Do you grab a screwdriver or a wrench?*"

Anyway, Truax and his Aerojet research team went over to NASA and explained all this stuff to the government men. "We took that into NASA and told them they'd been doing it wrong all this time. Of course, that's a bad thing to say to anybody."

The government men, predictably, didn't much like the sound

of this. They didn't like Big Dumb Boosters, or water launch, or water recovery, none of it. "After I developed the *Sea Dragon,* NASA did a 180-degree turn and opted for the much more complex winged system. If I'd have been in my grave, I'd have rolled over."

It was about this time that Truax got into the Snake River Canyon episode, but always in the back of his mind was his old *Sea Dragon* philosophy: water launch, water recovery, reusability. He wanted to prove this out someday, so when Evel Knievel appeared from out of the canyon and asked him what their next act was, Truax had his answer ready.

*F*inding volunteer astronauts was a lot easier than finding the cash. For both, Truax placed ads in the *Wall Street Journal:* "Wanted: risky capital for risky project." And: "Man or woman interested in becoming the world's first private astronaut—must be in reasonably good health and able to produce $100,000 in spendable money."

For a long time not much money poured in, but Truax was committed to his project to the point that he mortgaged his house to keep it going. After all, he had plans, drawings, dreams. And he had his surplus rocket parts.

He'd been walking through his favorite rocket-part junkyard in Ontario, California, one time when he spotted some Rocketdyne LR101 vernier engines, *seven* of them. Truax knew all about these engines. They were used for making course corrections on Atlas rockets after main engine shutdown, and to Truax they were works of art. The government had paid millions of dollars to make these things, and there they were, just sitting around rusting. Truax figured he could get them for twenty-five dollars apiece. "For twenty-five bucks," he said to himself, "I'll buy 'em, even if I have to use 'em for paperweights."

So Truax bought 'em. Later he yoked four of them together, to be the motive power behind the X-3, the "Volksrocket."

If surplus parts were easy to come by, so were astronaut candidates. In fact, Truax always had far more astronauts than he ever knew what to do with. There was Martin Yahn, for example, first in a long line, who at the time he volunteered happened to be unemployed and therefore unable to come up with the required $100,000. On top of that he was married and had two children. But he was nuts about going up into space, and whenever Truax rolled his Rocketdyne LR101 vernier engines out for static tests Martin Yahn would be there in his powder-blue jumpsuit marveling at the sights and sounds, enthralled. Truax was so impressed that he put Martin Yahn at the top of the list and decided to send him up for free.

But after a while Martin Yahn vanished into the mists of time, only to be replaced by others. One of them was Jeana Yeager, who was going to be the first *woman* private astronaut. Yeager was a resourceful worker and took on any challenge.

"One day Bob told me to build the launchpad for static testing," she recalled. "I drew up the plan and showed it to him. He calculated that it was just adequate to handle the thrust, suggested that I increase the dimensions slightly, and sent me on. We agreed on a suitable site, at Fremont Airport, and I hired the contractor and helped build the launchpad."

She also acted as a frogman in the water drop-and-recovery tests, where she'd jump into the San Francisco Bay and help maneuver the rocket so that it could be picked up by a crane. But finally Jeana Yeager too left the project, to work on someone *else*'s hubristic dream, which was to help design, build, and then copilot the first and only aircraft to fly around the world without refueling. That project suceeded in 1986, when she and Dick Rutan circled the globe aboard the *Voyager* on a single tank of gas.

Eventually, astronaut applicants started showing up at Truax's house with some folding money in their pockets. There was "Ramundo," stage manager for the Beach Boys. And there was Daniel J. Correa.

Dan Correa was from Peru. ("He's a bona fide Inca," Truax said.) The son of a mechanic in the Peruvian Air Force, and distantly related to a former president of the country, Correa and his wife arrived in the United States with about $150 between them.

"He heard about the X-3 project in the paper or something," Truax recalled, "and he came around to see me because he thought that his ancestors had come from outer space, and that it was his destiny to go back into outer space. He's a Rosicrucian, and they got some weird ideas."

Correa spoke Spanish and looked Mexican, and anyway he got a job in a tortilla factory rolling out the dough. Because he was always a very gung-ho, extremely ambitious type, he convinced the factory owner to put him out on the road selling tortillas on a commission basis.

Correa sold lots of tortillas, oceans of tortillas, so many you'd never think there were that many tortillas in the whole world, and after a while the owner was paying him off partially in the company's stock. Eventually Correa had acquired so much of the stock that he controlled, and then owned, the company, the Mission Bell Bakery, in Redwood City, California.

Then, right at the apex of his tortilla career, he decided to enter a hitherto unexploited market niche. The average housewife, he realized, had no good way of reheating frozen tortillas. If she put them in a frying pan they got greasy and burned before they were heated all the way through, whereas if she put them in the oven they dried out too fast and got brittle and ended up in a million pieces.

"So I redesigned my baby daughter's vaporizer and came up with this device for rejuvenating the tortilla."

It was Dan Correa's new invention, *The Tortilla Steamer*.

"The Tortilla Advisory Board is pleased with it," he said at the time, "and if I sell 350,000 steamers this year, I will make $5 million, plenty of money for the rocket."

Clearly, Dan Correa was Bob Truax's man. But like the Sky-Cycle X-2, the tortilla steamer concealed a tragic flaw that was not apparent at the very beginning. The steamer, which was a small box with a clear plastic top—it looked like a phonograph turntable—was an electrical can of worms. Steam would condense out on the top, drip down the outside, and get into the circuitry, where it would cause shorts and make a mess of everything. Unfortunately, before he submitted one of his steamers to the Underwriters Laboratory for its seal of approval, which it refused, Correa had already manufactured 10,000 units. He then had on his hands 9,999 non-UL-approved tortilla steamers.

What do you do with 9,999 non-UL-approved tortilla steamers?

Why, you ship them to Mexico, where consumers are not so uptight about having *seals of approval* on every last item, and you hope to God you can unload them down there.

By this time—it was early 1979—Correa had given Truax a healthy down payment on the rocket flight. "He got to $17,000 or $27,000," Truax recalled, "but then he ran out of money. He lost the bakery, he lost his house, and finally he lost what he had put into the project because he couldn't come through with any more. That was part of the deal, you know: if you didn't get the whole $100,000 then anything you put in was down the drain, because I was spending it as fast as he was putting it in. In fact I was spending it *faster* than he was putting it in! And so he lost the whole deal."

Correa returned to Peru for a while but was back in California again a few years later. "I'll get the rest of that hundred thousand," he told Truax. "I'm going to be back in this thing. Don't worry. We'll go."

On this occasion he brought with him another invention, a new type of building brick. These bricks were shaped in such a way that they interlocked with each other, so that they held together without benefit of cement. Somebody else had apparently invented these miracle bricks down in Peru, but Correa had managed to wangle a license to manufacture them in the States, where he now had visions of putting up vast tracts of mortarless houses.

But try as he might, Correa could not get anyone to mass-produce his hubristic bricks. He went to the San Jose Brick Company, which after careful consideration was forced to decline the honor. It would cost the company too much—in the hundreds of thousands of dollars, they estimated—to retool their machines. Truax surmised that even if someone could be persuaded to man-ufacture the brick, builders in this country would face insurmount-able problems with building-code regulations. So that was the end of Dan Correa.

When the astronaut position came open again, Truax was besieged with the usual nut-case phone calls. This was his own fault. He'd appeared on "The Tonight Show," and he was telling Johnny Carson and all the rest of the world about the X-3 private astronaut project, and Johnny seemed to love the idea, until Truax suggested that *he* be the victim. "I told Johnny he'd make a good astronaut," Truax said. "But he backed off."

Anyhow, people who wanted to be the World's First Private Astronaut were bugging the hell out of him ("I even had a *blind* guy who wanted to fly it!"), and at length he became a desperate man.

But then one night a San Jose businessman by the name of Fell Peters walked into Truax's garage and asked to go to the top of the list. "Well, it'll cost you $100,000," Truax told him.

Peters started laying $100 bills on the table, arranging them all into neat piles. Truax, who'd been through this kind of thing time and again (he'd sold the astronaut job four different times by then), expected a few thousand dollars to appear at most. But Peters was still going strong at $20,000. He kept on going even past $30,000.

Finally, the pile reached $40,000. Here, Truax admits, "I weakened." Fell Peters then went to the top of the list.

Later, Truax put Peters through his astronaut training program, which consisted of a ride in Truax's private plane, a Burt Rutan Vari Eze homebuilt. The ride included stalls, steep turns, and other hair-raising maneuvers, all to establish that the astronaut candidate could tolerate high levels of airborne stresses and strains. One of Truax's worst visions was that five seconds into the blast-off, which would be broadcast over live TV, the passenger would start screaming into the microphone, *"Let me out of here!"*

Bob Truax knew as well as anyone else how improbable the whole scheme was (just like the canyon shot had been, for that matter), but still he was utterly serious about private space travel. Sooner or later, he was sure, the X-3 really would lift up into the heavens with a live person aboard. It was no more than a right-thinking man could do with the proper combination of hubris, talent, and spare parts.

"We've got to stop thinking we're helpless," he said. "Hell, we knocked off the moon in ten years."

2

Home on Lagrange

F*in-de-siècle* hubristic mania was not by any means a new phenomenon. It had appeared on the scene at least once before, toward the end of the nineteenth century when, at about 1880, physicists decided that they had discovered virtually all there was to know about nature. That was when John Trowbridge, head of the Harvard University physics department, went around telling his students not to major in physics: every important discovery, he told them, had already been made. A few years later, in 1894, Albert Michelson, of the University of Chicago, announced that "the future truths of physics are to be looked for in the sixth place of decimals."

That was hubris. The very next year, 1895, Wilhelm Roentgen discovered X rays, and a few months after that Antoine-Henri Becquerel discovered the natural radioactivity of uranium. Suddenly it seemed that there was a whole new dimension to nature, and before the twentieth century was half over it became popularly known as "the atomic age."

To Jim Bennett, though, the century had an even more important mission, as became clear in the spring of 1976 when a physicist by the name of Gerard K. O'Neill came to the University of Michigan at Ann Arbor, where Bennett was a student, to deliver a lecture. Bennett had already read about O'Neill in *Time* magazine, so he knew that the physicist had a plan for putting these vast cities up into space, gigantic artificial habitats twenty miles long and four miles wide, containing up to ten million people each.

For Bennett, O'Neill's lecture rekindled an old interest, for he'd wanted to go up into space from the time he was nine years old and learned that Russia had just orbited the world's first artificial satellite, *Sputnik*. In the beginning it was just the romance of it all that attracted him, the idea of traveling to the moon or Mars. Later on that was coupled with a larger sense of purpose, as it seemed to Bennett that each century had its own historic objective: the nineteenth century had settled Australia and the American West, other centuries had other tasks, and he thought that the twentieth century's assignment was to conquer the last remaining frontier, outer space.

Bennett planned on being an astronaut but his eyesight was too poor for military flight training, and after a while it looked as if he'd never make it up to orbit after all. That wasn't such a tragedy, as it turned out, because by the time he got to high school the whole space program seemed to have evaporated anyway.

"I got kind of discouraged by the mid-sixties because I thought the space program was going incredibly slowly," he said. "Only four people on the moon by the end of the decade, and NASA was talking about a moon base with only a few dozen people by the seventies, and one or two piddly little Mars expeditions—to me this was nothing. It was not nearly enough, or quick enough."

So when he got to the University of Michigan, Bennett gave up on space and majored in the other subjects he had an interest in, political science and anthropology. He read a lot of history, about the past ages of exploration and migration, and was particularly fond of Samuel Eliot Morison's accounts of the great ocean voyages—Columbus, Drake, Magellan, and the like. And then one day Gerard O'Neill came to town.

O'Neill colonies wouldn't be "space stations," but rather small worlds: they'd be miniversions of earth, only instead of living on the outside of a big rock, as you did on the home planet, these space people would be living on the inside surface of a man-made habitat. Nevertheless, these artificial globes would be fitted out with all the comforts of home. They'd have everything up there anyone would ever need, including artificial gravity, housing developments, schools, hospitals, parks, lakes, streams, farms, skyscrapers, boats, bridges. Conceivably, there might be entire mountain ranges up there. It was as if you could take all of Manhattan island, plus a section of the Adirondacks, roll them up into a closed cylinder, and then float that cylinder along in a stable orbit between earth and the moon.

The idea of space settlements, Jim Bennett was aware, was not new. He'd been familiar with them from the science fiction books he'd been reading his whole life, where cities aloft were a regular element of the story line. The same general concept, in fact, had been proposed by others as far back as the turn of the century, in the last great wave of hubristic mania, when Tsiolkovsky wrote of the enormous space palaces that would be the future home of the human species. And then in his 1929 book *The World, the Flesh, and the Devil*, J. D. Bernal had gone a step further and imagined turning these habitats into rockets, into vast space arks that would travel off to the stars. Early twentieth-century technology was not quite up to the task, so all of it had to be dismissed as just so much pie-in-the-sky dreaming. O'Neill's proposals, by contrast, came at the very moment when the world's level of technological development had risen to the point that what he had in mind could in fact be accomplished. For it was all a matter of having the proper technology, a point O'Neill emphasized in his lecture by quoting plenty of hard data about how the colonies would be put together, where, when, and why. The man had gone into the concept in a way that none of his predecessors had, whether in fact or fiction.

So Bennett sat there and listened to all of the physicist's sensational claims about the political, aesthetic, and personal charms of living in self-contained orbital utopias. No one denied that his space habitats were possible. Rather, the question was whether they were *desirable,* whether they were worth the cost. O'Neill, at least, thought so. Space habitats, he said, would be "far more comfortable, productive, and attractive than is most of Earth." It was a thought that gave one pause; it took a while for it to sink in. Man-made worlds would be . . . *more comfortable, productive, and attractive than earth?*

But in fact this made a weird kind of sense, for who could doubt that the earth itself was, in its way, deeply flawed? The planet regularly suffered all kinds of natural disasters: volcanoes, earthquakes, tidal waves, droughts, floods, pestilences, and plagues. But artificial habitats would have *none* of those nuisances. You'd be able to *design them out*, you wouldn't be at the mercy of Mother Nature at all. Nature, in a sense, would be *gone,* for the whole space colony environment would be controlled, planned, regulated to the last degree. Industrial pollution would be a thing of the past: nothing that fouled up the atmosphere would be allowed in the living areas, no smoke-spewing factories, no smog, no automobile exhaust. It would be like heaven on earth—or close enough.

Best of all would be the politics of the matter. Space colonies, O'Neill believed, would be an unparalleled chance for humanity to break away from the authoritarian political structures of planet Earth. The habitats represented freedom, autonomy, escape from the creeping bureaucracy that seemed to threaten every conscious entity on terra firma. Space colonies, it seemed, had everything going for them and no immediately apparent drawbacks.

Such were Jim Bennett's thoughts as he listened to O'Neill's space-colony talk. So when the physicist mentioned that a group in Arizona had gotten together for the express purpose of publicizing his ideas and advancing the whole space colonization scenario, Bennett decided that he would have to go down there and take a look. It was called the L5 Society, and was run by a high-tech engineering couple by the name of Carolyn and Keith Henson.

Both Carolyn and Keith remembered quite clearly the first time they ever heard of Gerry O'Neill. It was a major turning point in their lives and, as it happened, took place in September of 1974, the very same week that Bob Truax was out in Idaho doing the Snake River Canyon shot.

The Hensons were engineering students at the University of Arizona at Tucson and were friends of Dan Jones, a researcher in the physics department and one of Keith's rock-climbing partners. One afternoon Jones came over to the Hensons' with the new issue of *Physics Today* in his hands. There was a great article in there, he told them, "The Colonization of Space," by Gerard K. O'Neill. It was an apparently sober proposal for putting a fleet of twenty-mile-long space colonies up in orbit.

This was so unprecedented for a major journal like *Physics Today* that the university's physicists stopped what they were doing to check out the author's facts and figures. "It shut down the department for the entire day," Jones told the Hensons. So far as the physicists could tell, though, there were no errors.

Carolyn and Keith were intrigued by the thought of space colonies; they were, after all, daughter and son of the space age. "We were married just before the first moon landing," Keith said later. "One thing we checked each other out on very carefully to make sure we had compatible attitudes was that when the first lunar base opened for colonization, we were going to be there."

Carolyn, a slim, lithe woman of considerable magnetism, might well have had space colonies in her genes. She was the daughter of Aden and Marjorie Meinel, both of whom were professional astronomers. Her father, who was once the director of the Kitt Peak National Observatory, a few miles south of Tucson, used to teach Carolyn all sorts of astronomical lore, leaving her as familiar with the solar system as most kids her age were with baseball. It was entirely natural for her to imagine herself rocketing out to the asteroids and setting up mining operations.

Keith was interested in science too, but he never cared much about knowing things for their own sake. Rather, he preferred to put his knowledge to practical use—as for example when he started Arizona's short-lived flying saucer industry.

He and his friends in Prescott would make these bogus craft out of plastic dry-cleaning bags, which they filled with hydrogen. Keith's father had an old aluminum-and-lye hydrogen generator in the family basement, and Keith would go down there, fill the bags with hydrogen gas, and send them up over the skies of Prescott.

During his college days in Tucson, Keith and some friends bought an unperforated two-thousand-foot roll of plastic tubing, cut it up into one-hundred-foot-long sections, and filled them with natural gas. They suspended homemade Japanese lanterns underneath, placed burning candles inside the lanterns, and sent the contraptions aloft. At about sundown these startling luminescent objects would start appearing all over Tucson: big, glowing lights in the sky floating around and making no sound whatsoever. What else could they be but *Flying Saucers?* The next day he'd read in the paper that several UFOs had been spotted the night before, manifestations that the air force officially attributed to "neon lights reflecting off low clouds."

Carolyn and Keith met at the University of Arizona. Later, in 1972, they started their own company, Analog Precision, Inc., to make automated devices for the mining industry. Mining was big business in Arizona and Keith got involved in the exploration end of it, going out into the field with company prospectors, who were always using dynamite for one thing or another. After a while Keith and Carolyn had become semiprofessional explosives experts.

"We were accomplished pyromaniacs," Carolyn said. "We were always going out in the desert and setting things off. Mostly just bombs."

That's to say, mostly just Recreational Explosives.

"Engineering students are always pulling pranks," she said, "but back in those days, before the SDS ruined it for us, you could just go down to the Apache Powder Company and buy whatever you wanted. You'd walk in and say, 'I want a case of dynamite, thirty feet of spritzer cord, ten yards of primacord,' and so on, and they'd sell it all to you like it was nothing."

So Carolyn and Keith took their dynamite and primacord and firing caps, and went out for the weekend "fire festivals," modest social gatherings where twenty or thirty of Tucson's finest young bomb blasters would compete to produce the biggest and best amateur detonations. Some of the visual effects out there were truly spectacular, like the time a couple of off-duty National Guardsmen came out with a gasoline-filled garbage can, which they then set into a basket of primacord, a linear explosive. This was supposed to be a mock atomic bomb, and indeed it worked pretty well.

"It made an *incredible* fireball and mushroom cloud," Carolyn said. "I mean, it was *really impressive.*"

Carolyn and Keith, though, had no trouble topping that one. They came back the next week with a device that would not only *look* like an A-bomb explosion, it would actually *work* like one.

The real atom bomb had an implosion lens detonator, and so too would Carolyn and Keith's. The core of the bomb would be a mixture of ammonium nitrate and diesel oil. They mixed this up easily in their garage; the only problem would be getting it to explode on command. "Ammonium nitrate is just hellishly difficult to set off unless you have tons of it together," Keith said. "You have to confine it and give it a pressure shock. It's a real pain to get it to blow up properly."

But *he* knew how to do it, of course, and so he took a two-hundred-pound lard can and put three pieces of primacord inside, looping them around so they completely covered the bottom. Then he poured the ammonium nitrate into the can, inserted sticks of dynamite all around the perimeter, and ran the primacord fuse up to a blasting cap on top of it all. The cap would fire the primacord, which in turn would set off the dynamite, which would crush the mass of ammonium nitrate until the necessary pressure was reached—a true implosion device, just like the atom bomb.

Carolyn and Keith thought that their new bomb would be *so* powerful that they took the precaution of putting it back behind a hillside, so that it would be out of anyone's direct line of sight.

So they set it all up, lit the fuse, ran like hell, and . . .

(Jumpin' hubris! This is how we learn about the forces of nature!)

. . . a shock wave blossomed forth like the world was coming to an end right then and there in southern Arizona. It was a misty, rainy day, and the ambient moisture condensed out in an expanding shell as the compression wave traveled out from the center.

It was stunning. Everyone agreed that it was a *very* loud explosion, one of the best recreational bombs they had ever seen, and the Hensons walked away easy winners at the fire festival that Sunday.

They and their friends also held "Ring parties," modeled after J. R. R. Tolkien's *Lord of the Rings.* "We reenacted the scenes that called for lots of fire, smoke, and explosives, like the storming of Isengard," Carolyn said. And as if all that weren't enough, the Hensons also owned a Civil War replica cannon named *Taras Bulba,* plus assorted guns, rifles, and other hardware. Indeed, Carolyn and Keith were every inch one of Tucson's highest-tech, highest-firepower married couples.

So in September of 1974, when their friend Dan Jones walked in with the space colony issue of *Physics Today,* Carolyn and Keith were no strangers to radical and weird ideas. O'Neill's article fit right in with their greater world view. "We probably made two thousand Xerox copies of it over the next five years," Keith said later. "I personally made five hundred. It was a self-replicating idea."

*T*he publication of the space colony article was a milestone for Gerry O'Neill as well. He'd been born in Brooklyn in 1927, the same year Lindbergh flew the Atlantic, and the story was that O'Neill's father took his infant son up to the roof of their brown-stone in Brooklyn and held the boy aloft so that he could see Lindy's return to New York. Later on O'Neill became a pilot, and in 1966 he even had a chance to go into space. That was the year NASA established the new category of scientist-astronaut.

O'Neill seemed to have everything going for him. He had the finest academic credentials (Swarthmore, Cornell, Princeton), he'd made a name for himself in experimental physics (he had developed storage rings for particle accelerators), and he was a sailplane pilot to boot. So he applied for a scientist-astronaut slot—he and about a thousand other people—and sure enough he made the final cut, becoming one of a group of sixty-eight scientists who went through the entire selection procedure. He went down to San Antonio, Texas, to the School of Aerospace Medicine at Brooks Air Force Base, for a week's worth of physical and mental tests. There he met some of the World-Historical Astronauts: Deke Slayton, Alan Shepard, and the rest. He went up in a T-38 jet trainer and flew aerobatics, and he loved it.

In all, he thought he had a pretty good chance, especially after he'd stood in front of the astronaut selection board and Deke Slayton looked him in the eye and asked, "Gerry, are you sure you're ready to give up everything you're doing back at Princeton? Are you sure you can be here in Houston, ready to go, on September first?" Oh yes, he was ready, all right.

It so happened that Alan Shepard was the man appointed to make *The Call,* the one that informed you whether or not you'd been selected into the astronaut corps, and it was his practice to begin each of these phone calls in exactly the same way. There'd be a long, formal preamble, and then he'd finally get to the real message—either you'd made it or you hadn't.

So one day O'Neill got *The Call*.

"Al Shepard here." And Shepard starts off with something like, "I know you've been waiting to hear of the committee's decision with respect to the selection of new astronauts. Well, the committee has now arrived at its decision and has forwarded the names to Washington," and blah blah blah—it seemed to take an eternity. But then came "unfortunately," and "I'm sorry to tell you," and it was clear all at once that that was the end of Gerry O'Neill's astronaut career.

Two years later, in the fall of 1969, Armstrong and Aldrin had walked on the moon and O'Neill was back teaching physics at Princeton. This was the time of the big turn against technology on the campuses, when students were asking *"So what?"* about the moon landing, and blaming science for everything from global thermonuclear holocaust (this was always just three minutes away, according to the *Bulletin of Atomic Scientists*) to bad karma, and O'Neill wanted to do what he could to stem the tide, to show that science and technology could address the important social issues of the day. So he formulated a question to give to his freshman physics class: "Is a planetary surface the right place for an expanding technological civilization?"

As to why he came up with that specific question, O'Neill himself never had much of a clue.

"There is no clear answer," he said, "except to say that my own interest in space as a field of human activity went back to my own childhood, and that I have always felt strongly a personal desire to be free of boundaries and regimentation."

But there was in fact a perfectly good reason why an imaginative physicist at the dawn of the space age would get around to asking whether the earth was really the best place for human beings. Indeed, who but the worst sort of Dr. Pangloss could possibly imagine that the earth was the best of all possible worlds, especially in the era of routine space flight, when a person could finally think about living elsewhere? The more you thought about it, the worse a place earth seemed to be.

At its most fundamental levels, the earth versus space colony issue was one of accident versus design: should we humans be satisfied with what we'd been handed as the result of accidental circumstances, or should we go out and apply some critical intelligence, some engineering skill, some good, old-fashioned hubristic science, to the problem?

As far as planet Earth was concerned, astronomical theory held that its very existence was a chance development. There was no reason at all why any given planet *had* to have happened; rather, the solar system came into existence when random clouds of interstellar gas and dust collapsed and condensed out to form large blobs of matter, and these blobs had become the sun and the planets.

Is that what anyone could call advance planning? The planets just clumped together and that was that.

The fact of the matter, as Gerry O'Neill and his students soon discovered, was that planetary surfaces were *not* the best places for human life at all. In fact, these surfaces were all wrong. The earth provided us with a place to live, but it did so at extremely great cost insofar as efficient use of materials went. The reason was that the earth was approximately spherical, and mathematically speaking a sphere was a minimal surface, meaning that it was the smallest area that's able to enclose a given amount of volume. Another way of saying this was that of all the different ways you might use a given amount of matter to create a surface, a sphere was the geometrical form that gave you the very *least* amount of area per unit of available mass.

A planet was the most profligate waste of raw materials imaginable. Underneath the surface of earth is lots of dead matter, some five thousand miles worth of dirt and rock—and what did any of it actually *do* other than slide the tectonic plates around (causing earthquakes), and spew out from time to time in noxious volcanic eruptions?

Of course, all this blobbed-together matter did one *other* thing too: it made for lots and lots of gravity. But you had to wonder whether gravity itself was all that wonderful a by-product. Certainly it was not necessary to human life, and in fact anyone could think of the needless inconveniences it caused: gravity made your face sag as you got older; it made earlobes and noses droop. And then there were the obstacles it made to just getting around: climbing up a flight of stairs, for example, was a major undertaking; shoveling snow a positive menace. The worst part about it was that there was no good *reason* for any of it: gravity was just there as an unavoidable artifact of all that matter.

If there was one thing that space science had demonstrated it was that life in zero G was far less strenuous on the heart (although admittedly it made the bones wither away) than life in the earth's one-G gravitational field. Not that there was anything special about *one-G* anyway: that too was another chance result, and the human species could have managed equally well, if not far better, with much less gravity. For that matter, humans could also have survived at even *higher* levels, as had been demonstrated repeatedly by experimental tests.

There was the hyper-G work done on chickens, for example, by Arthur Hamilton ("Milt") Smith in the 1970s. Milt Smith was a gravity specialist at the University of California at Davis who wanted to find out what would happen to humans if they lived in greater-than-normal G-forces. Naturally, he experimented on animals, and he decided that the animal that most closely resembled man for this specific purpose was the chicken. Chickens, after all, had a posture similar to man's: they walked upright on two legs, they had two non-load-bearing limbs (the wings), and so on. Anyway, Milt Smith and his assistants took a flock of chickens—hundreds of them, in fact—and put them into the two eighteen-foot-long centrifuges in the university's Chronic Acceleration Research Laboratory, as the place was called.

They spun those chickens up to two-and-a-half Gs and let them stay there for a good while. In fact, they left them spinning like that day and night, for three to six months or more at a time. The hens went around and around, they clucked and they cackled and they laid their eggs, and as far as those chickens were concerned that was what ordinary life was like: a steady pull of two-and-a-half Gs. Some of those chickens spent the larger portion of their lifetimes in that goddamn accelerator.

Well, it was easy to predict what would happen. Their bones would get stronger and their muscles would get bigger—because they had all that extra gravity to work against. A total of twenty-three generations of hens was spun around like this and the same thing happened every time. When the accelerator was turned off, out walked . . . *Great Mambo Chicken!*

These chronically accelerated fowl were paragons of brute strength and endurance. They'd lost excess body fat, their hearts were pumping out greater-than-normal volumes of blood, and their extensor muscles were bigger than ever. In consequence of all this, the high-G chickens had developed a three-fold increase in their ability to do work, as measured by wingbeating exercises and treadmill tests.

So they stomped around on the treadmills and flapped their wings and they proved to one and all that, yes, indeed, here was a fabulous new brand of chicken. But the question was, *What was it good for?* It wasn't as if all that extra blood, bone, and muscle was exactly *needed* for anything—not in the one-G field of normal and everyday life on earth.

On the other hand, an apologist for earth might claim that there *were* some benefits we derived from our one G: for example, gravity was what held the atmosphere onto the earth's surface. But from the engineering point of view, that was a remarkably unintelligent way of arranging things. It was like a bank storing its money on the outside, hoping to keep it there by magnetism or something of the sort, instead of sealing it up in a vault. The final payoff of Mother Nature's misguided atmosphere-on-the-outside policy was nowhere better illustrated than on Mars, where most of the atmosphere had long since leaked away into space.

Besides all that, earth's gravity brought with it a special drawback in the space age: it posed a major hurdle to space flight. Blasting out of our gravity well, Gerry O'Neill once said, is equivalent to "climbing out of a hole four thousand miles deep, a distance more than six hundred times the height of Mount Everest."

You had to climb six hundred Mount Everests just to get out and free of earth's gravity, circumstance that led O'Neill to speak of human beings as "gravitationally disadvantaged." That's what all that gratuitous dead matter down there inside the earth did for you: it made you *Gravitationally Disadvantaged.*

But gravity was not the *only* drawback of living on planetary surfaces; there were plenty of *others* as well, such as the day-night cycle that was a product of a planet's rotation, for example. However useful it might be for sleeping, night was of absolutely no benefit to plants, which could grow all the better in twenty-four-hour-a-day sunlight. But of course you couldn't *have* twenty-four-hour-a-day sunlight with that dead planetary mass hanging out there blocking sun rays. A properly designed space habitat, by contrast—an *intelligently* designed one, as opposed to one that developed *by accident,* here on earth—would allow a constant stream of sunlight to penetrate down to crops.

Nor would there be any *seasons* to worry about in a space colony. The seasons resulted from the tilt of the earth's axis, something *else* that emerged by accident and was really wholly unnecessary. There was no good reason for it, especially when you considered what it did to agriculture, which is to say that it made it next to impossible for a large part of the calendar year. *None* of this stuff would be tolerated in an artificial habitat: if seasons were permitted to exist at all they'd be equally helpful to animal and plant.

From every enlightened engineering standpoint, then, it was all too obvious to Gerry O'Neill and his Princeton students that planet Earth was *not* the best of all possible worlds, that in fact going the planetary route was probably the *worst* way to go if what you were concerned with was making maximum use of resources, optimizing agricultural yield, getting out into space, and distributing the population into smaller, autonomous communities, thereby promoting human welfare and freedom.

The interior surface of a hollowed-out enclosure, they decided, would make for a better living environment than the outside surface of a planet. First of all, the atmosphere would be kept in by the surface, as in a bank vault. You could rotate the space colony to provide gravity, which would vary according to location. Activities that were better in low-G—like hang-gliding, swimming, or sex—could be practiced in the low-G zones, while higher-G areas might be reserved for living, manufacturing, or agriculture. The seasons would be controlled, as well as the weather, and plants would have constant sunlight, for there would be windows on all sides through which sunlight could be angled in, by means of mirrors if necessary.

Soon O'Neill was envisioning complete new earths and describing the contours of mostly artificial but nevertheless fantastically lush places in the sky, in words that would have to be dismissed as the most atrocious crackpottery were it not for the fact that there was no reason to doubt that any of it was possible.

"I would have a preference, I think, for one rather appealing arrangement: to leave the valleys free for small villages, forests, and parks, to have lakes in the valley ends, at the foot of the mountains, and to have small cities rising into the foothills from the lakeshores. Even at the high-population density that might characterize an early habitat, that arrangement would seem rather pleasant: a house in a small village where life could be relaxed and children could be raised with room to play; and just five or ten miles away, a small city, with a population somewhat smaller than San Francisco's, to which one could go to theaters, museums, and concerts."

Rather heady stuff, to be sure, but it was the first flowering of *fin-de-siècle* hubristic mania there at the dawning of the space age, at the point where science and technology had caught up to the most excessive dreams of fiction.

In retrospect, it was inevitable. Why *not* think about reengineering the earth—and about getting it *right* this time, or at least putting it all together more *intelligently,* with planning, design, forethought, and conscious intention? Enough of this living on the accidental remains of a collapsed interstellar dust cloud. We've seen the earth and all its flaws, all its drawbacks, and now we've seen the outlines of something better. And since we've seen all that, and we have the wherewithal, technologically and scientifically, to do it, well then, why shouldn't we go ahead?

*T*his was exactly the type of reasoning that appealed mightily to Carolyn and Keith Henson. And how understandable this was. There they were, a couple of extremely intelligent engineering types—an astronomer's daughter and a guy who could take the world apart with a screwdriver—living in the Wild West, in Tucson, Arizona, where it's still, to this day, absolutely legal for an average person to carry a loaded revolver in plain sight, as in a holster. There they were, a high-tech, high-firepower couple who spent their weekends setting off bombs out in the desert, both of them science fiction fans since just about the time they could read; there it was, the dawn of the space age, and into their living room walks physicist Dan Jones with Gerry O'Neill's blueprint for a celestial city in his hands. What were they supposed to do, sit back and laugh their heads off?

"I absolutely wanted to go into space," Carolyn said. "I wanted to live there and grow food. I wanted to be a pioneer, in the classic spirit."

"There really isn't much left to do here," said Keith. "The highest mountains and the lowest valleys have all been explored on earth. The opportunities are rather limited."

"In other words," Carolyn said, "we were worried about things getting very, very BORING if we stuck around on this planet too long."

The only frontier left was the one overhead, and there were already people up there—astronaut Al Shepard had hit a golf ball on the moon, astronaut Ed Mitchell had tried out ESP up in orbit—so what could be more natural than for human beings to *get on with it?*

Science fiction authors had portrayed in vivid detail what a great adventure it would be to live elsewhere. The first such account that Keith ever read was Robert A. Heinlein's *Farmer in the Sky*. It was a children's book, written for the Boy Scouts of America, so scouting was a recurrent theme, but the story was about colonizing the solar system, about a boy and his family going off to Ganymede, one of Jupiter's satellites, to escape overpopulation and famine on earth. Ganymede had already been given an artificial atmosphere by the use of "mass-energy converters," which turned ice into energy and atmosphere.

"The material was there—ice," the narrative said. "Apply enough power, bust up the water molecule into hydrogen and oxygen. The hydrogen goes up—naturally—and the oxygen sits on the surface where you can breathe it."

According to the story an earthling could come up to Ganymede, homestead twenty acres of land, and then "at the end of five earth years he owns a tidy little farm, free and clear." So Boy Scout Bill went up there with his family, and they cultivated their tidy little farm. They survived all sorts of calamities, including failure of Ganymede's "heat trap," the man-made device that regulated the weather. At the very worst moment, when it seemed that a new ice age was about to overtake the planet, Bill and his family were about to call it quits and head for home, but at the last minute they decided to stay. There were, after all, compensatory benefits.

"A harvest moon looks big, doesn't it?" says Bill's father. "Well, Jupiter from Ganymede is sixteen or seventeen times as wide as the moon looks and it covers better than two hundred and fifty times as much sky. It hangs there in the sky, never rising, never setting, and you wonder what holds it up. . . . It's a show you never get tired of. Earth's sky is *dull*."

Well, this is what Keith Henson believed, all right, and by the time he came to read Gerry O'Neill's space colony paper, you might say he was primed for the idea.

At the time he got Keith Henson's first fan letter, O'Neill was in the process of organizing the second Princeton conference on space colonization. The first one had taken place on May 10, 1974, three months before his *Physics Today* piece was published. The first conference had been a success, having attracted about 150 people, including Eric Drexler, an MIT undergrad who helped organize it; astronaut Joe Allen; physicist Freeman Dyson from the Institute for Advanced Study; and people from NASA, the sciences, and the aerospace industry. The media was there as well, and on the Monday after the conference there was a front-page story about it in the *New York Times*.

One of the topics to be covered at the next Princeton conference, to be held in May of 1975, was agriculture in space. This had become a pressing matter to potential Space Rangers ever since NASA, in its typical institutional fashion, had decided that for technical reasons the optimum diet for long stays in space should consist of such items as algae, yeast, and distilled urine.

"This was some kind of engineer's perspective," Eric Drexler said later. "You know, 'Let's feed them some homogeneous liquid stuff that we can pump around easily.'"

How fortunate it was for the future of space colonization, then, that when he'd written to O'Neill, Keith Henson had mentioned his family's expertise in farming. "I talked about our bicycling to work, gardening and raising a lot of our own citrus fruits, raising rabbits and goats and chickens, all on a very small lot in the middle of a large city—what I call an integrated life style. Because it seemed to me it would be applicable to living in some sort of space colony, where you have the problem of doing a lot with a small area."

O'Neill was so impressed with their "integrated life style" that he placed a phone call to the Hensons and asked if they'd come up to Princeton and give a paper on space agriculture at the next conference, all expenses paid. It was an offer they couldn't refuse, so at the appointed hour Carolyn and Keith were giving their "Space Farm" talk up there onstage.

Their point was that if you *had* to make a lot out of a small farm in space, you could certainly do so. There were the techniques of multiple cropping and interplanting, for example, methods that had hardly been tried on earth—indeed, most farmers had never even heard of them—but when they *were* tried out, they produced sensational results.

Interplanting was the technique of sowing the seed for the next crop before the first one had been gathered up. Usually a farmer waited until a cornfield, for example, was harvested before planting anything else in the same field, but if you planted the next crop ahead of time you'd give it a head start and thereby get a bigger yield. Multiple cropping made the output even larger. Usually a farmer planted a crop in parallel rows but put nothing in the open spaces between the rows. This, though, was just a waste of precious soil. Why not plant some low-growing vegetable, such as sweet potatoes, in between the rows of a high-growing one, such as corn? It would be like getting something for nothing, two crops for the price of one.

This was not just wild theorizing, for both these techniques had been tested at the International Agricultural Experimental Station in the Philippines, with the result of more than quadrupling the average per acre output of farmland. The yield would be even better than this in a space colony, where there'd be constant sunlight and perfect growing conditions, with controlled lighting, air composition, and temperature. Indeed, the University of Arizona's

Environmental Research Lab often got yields twenty times those of the average farmer.

As the Hensons envisioned it, the space farm would also include livestock, principally rabbits. Carolyn and Keith raised rabbits on their small plot in Tucson and knew from experience how efficient the animals were in converting feed to meat.

At the very least, space colonists would not be reduced to yeast cakes and algae-meal. "Space would be a land of milk and honey," Carolyn said, "of french fries and rabbitburgers."

Later, when she attended the NASA Summer Study at the Ames Research Center, in Mountain View, California, Carolyn wanted to demonstrate the culinary advantages of her rabbits-in-space project.

"I don't like the idea of everything being totally intellectual," she said. "I figured, why argue about the edibility of rabbit when you can simply feed it to people and let them make up their own minds."

So she and Keith took about ten rabbits from the farm, slaughtered them, stuffed the meat into plastic bags, and got on the plane to San Francisco. A few hours later they were at the Summer Study, where Carolyn served up her standard banquet of fried rabbit, jalapeño peppers, and homemade goat cheese. There were a few jokes about "rabbit-propelled space colonies," and so on, but everyone had to admit that rabbit meat tasted a lot better than algae.

A new social order was soon springing up among those who had attended O'Neill's Princeton conferences and the Ames Summer Study groups. And why not? While everyone else was back home watching baseball and getting fat on tacos and beer, these people were out there plotting the next major step in human evolution. Space colony songs were written and sung: "This Is the High Frontier," by Tom Heppenheimer, and "Reach for the Stars," with lyrics and music by Carolyn Henson:

> We gather here together to create the future of Earth,
> We are joined together, humankind in rebirth.
> The universe is open, the gates of the stars open wide,

Lands of milk and honey in the starry fields of the sky.
The universe is open, the future rests in our arms,
Reach for the cosmos, reach for the stars.

Newsletters were planned, sign-up sheets passed around. There was talk of putting together a "High-Frontier Society," or something of the sort, a kind of gung-ho space-activist group. Naturally, the task of organizing it fell to the most vocal and visible among them, which is to say Carolyn and Keith Henson.

Back in Arizona from the first Princeton conference the Hensons incorporated something called the "L5 Society," named after the fifth Lagrangian region between the earth and the moon. These five regions had been postulated by the astronomer Joseph-Louis Lagrange, back in the 1700s, as locations where the gravitational attraction between the earth and its satellite would cancel each other out. If a third, much smaller body were placed in one of those regions it would tend to stay where it was all by itself.

Gerry O'Neill and friends had decided that this is where the first space colonies should go. Indeed, O'Neill had written of eventually putting "five thousand habitats at each Lagrange region," enough to house a total human population many times that of earth.

In September of 1975, four months after the second Princeton conference and a year after O'Neill's original *Physics Today* paper that started it all, the first issue of *L5 News* was published from out of a back room at the Hensons' electronics equipment firm, Analog Precision. The L5 Society, the newsletter said, had been formed "to educate the public about the benefits of space communities and manufacturing facilities, to serve as a clearinghouse for information and news in this fast-developing area, and to raise funds to support work on these concepts where public money is not available or is inappropriate. . . . Our clearly stated long-range goal will be to disband the society in a mass meeting at L5."

Early members included Eric Drexler, Hans Moravec, Saul Kent, Timothy Leary, and Marvin Minsky. Isaac Asimov and Robert Heinlein came on as directors, as did Freeman Dyson and Jerry Pournelle. Eventually the society grew to about ten thousand members, and after a while it was as if you couldn't meet a forward-looking thinker who wasn't a card-carrying fan of the L5 Society and who couldn't quote you his or her membership number, the lower the better. ("Mine's 363," Hans Moravec once boasted.)

A low membership number was not, however, the supreme status symbol among the L5 Space Rangers. The supreme distinction was to have visited the Hensons at their office or, preferably, at home. Ostensibly you'd do this to "help out"—with the office work and such—but that was not the *real* reason for making the trek down to Tucson. The real reason was to meet this couple, this high-firepower Carolyn and Keith, Mama and Papa Space Ranger. It was as if palpable lines of attraction emanated forth from this stellar pair, and you just had to go down there and see it for yourself, experience it, and bask in the glow.

One of these pilgrims was Jim Bennett. The Hensons, he discovered, indeed had ample farming experience. You expected them to be living in a proto-starship command center, but in fact they lived in a suburban ranch house that had a big garden out back plus a fenced-in collection of Old MacDonald barnyard animals. The Hensons kept goats, chickens, turkeys, rabbits, pheasants, pet peacocks, what have you. Most of the animals were kept as practical food items, not for aesthetics or zoo value, although a few of them—the goats—were treated as pets, and had been given the names "Popcorn" and "Art Planet Earth," the latter of which had been named by the Henson kids, Windy and Gale, as a pun on "*Our* Planet Earth."

Anyone had to admit that the goats made for some comic relief, as for example when one of them would happen to wander into the kitchen and butt you in the rear end while you were pouring raw goat's milk over your hydroponic granola or whatever it was. Then there was the time that a neighbor called up and complained that one of the goats—Art Planet Earth, as it turned out—had wandered into their house and had gotten up on the kitchen table, where it was doing some weird kind of goat dance.

And there were all these antic stories and rumors, like the one about the New York City lawyer who came down to see the Hensons and had his tie eaten by a goat. ("That's impossible," Carolyn remarked much later. "A goat might swallow a tie, but they don't have opposing incisors, so they can't actually bite anything off.") Jim Bennett swears to this day, though, that he personally saw the goats eating some papers that a visitor had tucked carelessly under his arm.

The Hensons spent their evenings at home with the kids, and for entertainment they'd fool around with the tesla coil that they had in place of a television set. "Keith's idea of a fun evening," Jim Bennett said, "was sitting around the tesla coil lighting up neon tubes."

A tesla coil is a device for generating high-frequency, high-voltage currents and had always been favored by science and engineering types because it let you command one of the basic forces of nature, electricity. Bob Truax built a tesla coil when he was a kid and used to produce two-foot-long arcs of lightning with it. Keith Henson put together his own model while he was in high school, and the device generated a million volts. When it was running it would throw these gigantic electrical arcs from ceiling to floor, and would light up any fluorescent lamp in the immediate vicinity. It also disrupted TV reception all over the neighborhood.

Another prime entertainment *chez* Henson was finding your way through the maze of tunnels underneath the house. These had been dug by a neighbor of theirs named Wizard, also known as The Human Mole.

"He dug two hundred feet of tunnels under the house," Keith Henson recalled, "and one of them came up in the kids' bedroom closet. I remember one night there were about ten people over here visiting, and there was this hole, about eighteen inches by fourteen inches, that had been sawed out of the concrete slab in the bedroom, and it went down into this network of tunnels. And so this whole batch of people, they all went down into this tunnel system and disappeared. There was an outside exit and they got out that way. The kids thought this was the greatest thing in the world."

As to why Wizard ever dug all those tunnels in the first place . . . "He just enjoyed digging," Henson said.

Timothy Leary also went down to see the Hensons. "I visited them several times," Leary said. "They had this extraordinary maze of cavernous runways into their house. In every way they were a science fiction couple."

Leary, as it turned out, would become an extremely committed

L5 member, recruiting about a thousand new parishioners and writing several articles for the *L5 News* in his own special brand of Mentally-Expanded Prose: "Space Migration offers our unfinished species the opportunity to create new realities, new habitats, new neural perspectives, new worlds unlimited by territorial longitudes or gravitational chauvinisms."

He'd first gotten serious about space travel while he was in Folsom Prison, serving time on drug charges. "In those circumstances one's thoughts and fantasies tend to lead to notions of escaping gravity. Watching sea gulls in the yard of Folsom Prison, for example, leads the eyes upward."

Leary remembered a meeting with Carl Sagan, who, being in favor of unmanned, automated space probes, was not an L5 member. "There was a scientific conference in San Francisco," Leary said, "and through mutual friends I invited Sagan to come out and visit me, and he did. I had chains on my wrists—I was in a special lockup thing called 'the hole' at that time; they wanted to keep me away from the other prisoners because they thought I was having too much influence on them—and anyway Sagan asks me, *'Timothy, why are you so interested in going out into space?'* and I looked at him and rattled the chains, and I said, *'Are you kidding?'*"

As dedicated as he was to space colonization, Leary drew the line at meeting Gerry O'Neill: he didn't want to embarrass the man. "He had *enough* trouble selling space to Congress," Leary said. "Can you imagine: *'Oh, here's Dr. Leary!' 'Well, that's pretty spacey!' 'Why don't you put your beanie on, Doctor, and tell us what you're gonna smoke to get up into space?'*"

But Timothy Leary was not, by any means, the most iconoclastic L5 member to come down and visit Carolyn and Keith. Some of the more exotic specimens would show up at the Hensons with their entire wordly possessions in a backpack and announce that they'd come to live with the couple for the rest of their lives. Carolyn remembers having to get rid of one particularly offensive fellow who wanted her to turn his pool table into a spaceship.

"Mostly the nuts would write or call up," Jim Bennett recalled. "The Hensons kept a nut file in the office, filled with the crankiest and craziest letters. There was this one guy who was trying to build a starship—a true interstellar starship—in his front yard in northern California. People would write in with the antigravity theories, the perpetual motion theories, long diatribes about how the government is suppressing this or that secret information. Someone was trying to insist that the government had a base on Mars since 1964, that flying saucers were actually secret Pentagon spaceships captured from the Germans during World War Two, that Martin Bormann was flying UFOs out of bases in South America. You get the idea."

Back in Princeton, meanwhile, Gerry O'Neill was developing the hardware necessary for building space colonies. His scenario was to start from earth, lifting the bare minimum of structural components, supplies, and personnel up into space by means of ordinary chemical rockets, and then as soon as possible thereafter start hauling raw materials from the lunar surface itself. There were two main reasons for this: the moon's lesser gravity, and its lack of atmosphere. The absence of atmospheric resistance on the moon would allow use of a new kind of launch device, the mass-driver.

The mass-driver was basically an electric cannon, a long tube from which batches of lunar ore would be shot into space on measured bursts of electrical energy. The ore would be caught by a mass-catcher—a quarter-mile-long funnel hanging in space—and then transported to orbital factories, where it would be processed and fashioned into the beams, girders, or whatever other structural components were needed to make the space colonies themselves.

The idea of such a launch system did not originate with O'Neill and in fact had been proposed by a number of others, such as Arthur C. Clarke, who in a 1950 technical article proposed using electromagnetic launch from the moon. Even earlier, in 1937, a Princeton University physicist by the name of Edwin Northrup had described a similar scheme in a science fiction novel called *Zero to Eighty*. But O'Neill now revived the idea and, together with MIT researchers Henry Kolm, Eric Drexler, and others, made some small, proof-of-concept test models. The prototypes worked so well in the lab that there was not the slightest doubt in anyone's mind that full-size models would work as advertised on the lunar surface.

But all this assumed that we'd already gotten up to orbit, whereas the initial hurdle was just getting out there into space to begin with, hauling building materials and a start-up population of gravitationally disadvantaged human beings up out of the earth's four-thousand-mile gravity well. This was the nub of the problem, and there was no lack of ideas for overcoming it. The main division of thought was whether to rely on government or on private enterprise to accomplish the task. Not that anyone cared that much either way—the main thing was just to *get out there*—but it turned out that few L5 members had any confidence that government spacecraft would be up to the task. This was particularly ironic in light of the fact that no human agency other than the world's biggest governments had yet succeeded in getting any piece of hardware into orbit. Nevertheless, there was all kinds of hostility toward NASA.

For one thing, L5 members were fed up with NASA's policy of obsessive perfectionism, which meant that every launch had to be perfect, which meant triple backup systems and all manner of superfluous redundancy, which meant an inordinate amount of testing and expense. Second, NASA tended to operate bureaucratically—witness the algae food business—and favored military as opposed to peaceful or scientific payloads. And of course NASA's penchant for sending up only one type of physical specimen meant that about half of the L5 membership wouldn't have the ghost of a chance of getting up there. NASA's bias in favor of straitlaced types was especially offensive to Timothy Leary.

Leary was annoyed by a famous statement of Thomas Paine's, then the director of NASA, to the effect that the moon landing was "a victory for the crew-cut guys who use slide rules, read the Bible, and salute the flag." Leary wrote in the *L5 News* that Paine, "in one magnificent flight of chauvinist rhetoric, manages to alienate all taxpayers who are female, non-engineers, non-Protestants, and prefer non-Marine hairstyles. Substitute the word '*Marx*' for '*Bible*' and the NASA version of whose reality should control space is in close agreement with Soviet planners."

There was thus understandable preference among L5 members for private launch vehicles, especially for the variety of off-the-shelf, high-risk hardware as had been advocated by Bob Truax. So it was encouraging news to L5 members when in May of 1977, at about the same time that O'Neill was demonstrating his Model I mass-driver at the third Princeton conference, a German launch-vehicle firm successfully fired off the world's first private rocket meant for commercial use. The company was called OTRAG.

OTRAG, short for *Orbital Transport-und Raketen-Aktiengesellschaft*, was based in Munich but operated out of its own private principality within the African state of Zaire. This by itself was an impressive accomplishment: the launch site was in fact a 39,000-square-mile territory, an area about as big as the state of Virginia. The company had struck a deal with Zaire's President Mobutu to lease this tract of land—exactly as you might lease an office or apartment—for twenty-five years, beginning in 1976.

It was as if they had been given permission to start up their own private government, for inside this territory the Germans were made exempt from Zairian law, and were given the authority to establish their own rules, regulations, and judicial system. They were also granted the right to banish from the area anyone they wished—including even the ten thousand native bushmen who already lived there. The rent for all this was $50 million per annum, but the company expected to make this and much more from its private launch service. It would be fly-for-hire, more or less like a privately run airline, except that instead of going from city to city, your cargo would end up in orbit. "OTRAG will launch reconnaissance satellites and earth resources satellites and others too dangerous or politically sensitive for shuttle launch," said the company founder, Lutz Kayser.

The launch complex itself was up on the Manono plateau, a four-thousand-foot-high plain overlooking the Luvua River Valley. Living conditions up there were primitive but pleasant enough, for the Germans had taken a tract of jungle inhabited mainly by snakes, bats, and wild pigs, and turned it into a kind of summer camp. The scientists lived in a colony of thatched huts scattered through stands of shade trees. Nearby were a dining hall and recreation center; an outdoor swimming pool was still in the planning stages.

Off in a separate area a seven-thousand-foot airstrip had been gouged out of the jungle for the four-engine Argosy transport aircraft that shuttled in and out of Munich with the compound's food supplies, machinery, and rocket propellant. The launchpad itself was away in a clearing toward the edge of the plateau, where a couple of tree trunks had been lashed together and joined at the top by a crossbeam. It looked like a gallows, but was in fact a rocket gantry.

Primeval as it all was, it worked well enough, for on May 17, 1977, OTRAG's first launch vehicle—a slender, thirty-foot, liquid-fueled craft—reached the end of its countdown and ignited. Instead of exploding in the conventional manner, the rocket rose up smartly and continued on straight as an arrow for the next twelve miles.

The entire flight was a testimony to the Bob Truax style of off-the-shelf rocket philosophy. OTRAG had been founded by a private-enterprise Wernher von Braun. An early bloomer so far as rockets were concerned, Lutz Kayser had, while still in high school, formed a rocket club, the German Society for Rocket Technology and Space Travel, and then, before he even graduated from high school, had invented what was then the world's smallest liquid-fueled rocket, the TIROC, for "tiny rocket." About as big as a shot glass, it generated half a pound of thrust, enough to make small vector adjustments on orbiting satellites. By the time he was firing launch vehicles out of his own personal Cape Canaveral in Africa, Kayser had patented the TIROC along with some thirty other space hardware designs.

During the 1960s Kayser and two colleagues, Eugene Sanger and Wolf Piltz, came up with an idea for a "space truck," a simple, cheap, workhorse rocket. There had been lots of ideas for "space trucks"—von Braun himself had proposed shuttlelike designs as far back as 1952—but the problem with all of them was that they were government craft, which meant that the designers apparently had to exercise no cost control measures whatsoever. It wasn't like you were operating a business, where costs had to be tailored to profits, and Kayser had a particular fondness for criticizing NASA approaches to space hardware. "Did you know," he'd ask visitors, "that NASA allocated about half a million dollars just to develop a ballpoint pen for the Apollo missions that would work under conditions of weightlessness when, after all, an ordinary pencil would have done the trick just as well?"

The same kind of wastefulness infected designs for the "simple space trucks" that were then on NASA drawing boards. The designers loved exotic new technology, and would cram all sorts of complex mechanical wonders into their spacecraft, and just in case one of these new marvels should happen to suffer an occasional performance anomaly (i.e., it failed), there would have to be a backup, which meant one or more redundant systems, so that by the time they got finished with it their simple "space truck" had been transformed into the *Queen Mary* of the cosmos.

To Lutz Kayser, all this was totally unnecessary. To him, the very idea of using a manned vehicle—the space shuttle—for the purpose of placing satellites into orbit made no sense. His policy was just the opposite: use strictly unmanned vehicles for cargo, cut corners wherever possible, build the rocket out of mass-produced, interchangeable parts, use assembly line techniques, and so on. It was the Bob Truax philosophy every step of the way.

Kayser's own rocket was based on the building-block concept. You started with a tank, engine, and guidance system: that was the basic unit. To get more thrust, for a heavier payload or a higher orbit or whatever, you didn't go back and design a bigger vehicle, all you did was cluster together more of these same building-block units until they collectively added up to the thrust you needed. You just strapped them together, like so many stalks of asparagus.

As Kayser envisioned it, the minimum controllable unit would be a cluster of four building-block units. Steering would be accomplished not by gimbaling the engine, which is to say tilting it back and forth in its mounts, but by varying the power outputs of the four separate engines, each of which could be controlled individually. Throttling back one of the engines would cause the whole vehicle to pitch over one way; throttling back another one would angle it over the other way, and so on.

For bigger payloads, whole clusters of these building blocks would be strapped together and fired sequentially, the outermost ones first. The outer rockets would fall off as they were used up, then the next layer in would take over. The higher it went, the thinner the launch vehicle would get, until finally only a slender core module was left to boost the payload into orbit.

Each building-block unit was an assemblage of off-the-shelf parts, few of which had ever been designed for spaceflight. The body of the rocket was industrial pipeline. The engine's ball valves were stock items commonly used in chemical plants, and the valve inlets would be controlled by Volkswagen windshield wiper motors. The propellants—kerosene and white fuming nitric acid— would be fed into the engines not by means of the usual turbo-pumps, which were expensive, but by the so-called blow-down method, which was simplicity itself: compressed air inside the fuel tanks kept the propellant flowing out, just like in an aerosol can. It was about as basic as a rocket could be and still work.

And it did work. The first successful flight, the May 1977 lift-off from Zaire, was with the minimum controllable unit, a four-rocket module, and it went up without a hitch. OTRAG did all this not with the intention of being a space colony pioneer, but simply to make money. It would sell its services to anyone who'd pay the price, American or Russian, capitalist or communist.

"For a president of a company who is responsible to its share-holders," Lutz Kayser said, "it is not an easy matter to refuse a customer." They even had a few customers lined up, including the government of India, which wanted to orbit its own TV satellite. According to Kayser, the Indians thought that if they could get a television set into every village for nightly entertainment then their population-growth problem would be cut in half.

OTRAG never flew any satellites for India, but the company's activities were always well covered by the *L5 News,* in a series of illustrated articles.

Who could tell? Maybe OTRAG would lead the way into space.

*S*uddenly, after OTRAG, private launch vehicles became a new cottage industry, a sponge for excess venture capital, so commercial rocket companies were springing up all over the place. And why not? After all, this was the way science fiction had said it would go. In Jules Verne's *From the Earth to the Moon,* for example, the world's first moon rocket was sent up not by any government agency but by the Baltimore Gun Club, a group of pyromaniacs who would have been right at home at Carolyn and Keith's fire festivals.

A few years after he first saw Gerry O'Neill in Ann Arbor, Jim Bennett moved to California, where he got involved with private launch vehicles himself. Out in California Bennett worked for a while for Gary Hudson, a self-taught rocket engineer from Minnesota. Hudson had designed a rocket called the *Percheron,* named after the French draft horse. Supposedly, the *Percheron* and a succession of follow-up vehicles would take anyone who could pay for it up into space, earning the inventors a large fortune in the process. Unlike the OTRAG prototype, however, the *Percheron* behaved more conventionally, which is to say that when the countdown reached zero it exploded on the test stand.

Fortunately for Bennett, he'd quit Gary Hudson's company a few months earlier. He then worked for a couple of others: Space Enterprise Consultants, and ARC Technologies, which later became Starstruck. Starstruck launched one rocket on a fifteen-second test flight and then went out of business.

Finally Bennett met George Koopman. Koopman, a good friend of Tim Leary's, had been an early backer of the L5 Society. Like Bennett, who never finished at the University of Michigan (they gave no courses in rocketry, after all), Koopman was a college dropout. He joined the army, where he made training films, then went on to Hollywood, where he became a stunt producer. He worked on *The Blues Brothers,* the movie starring Dan Aykroyd and John Belushi, and managed to talk the Federal Aviation Administration into letting him drop a Ford Pinto from a helicopter hovering at fifteen hundred feet into a river in the middle of Chicago. At first the Feds said no, but Koopman did a test drop in an Illinois cornfield and convinced the FAA that no harm would result. So they gave permission, and the stunt finally appeared in the movie.

Later, Koopman, Jim Bennett, and an engineer by the name of Bevin McKinney designed an entirely new type of launch vehicle, one that absolutely *could not explode,* and the three of them founded the American Rocket Company (AMROC), of Camarillo, California, to manufacture it.

The garden-variety rocket used liquid propellants, such as kerosene and liquid oxygen, which were typically stored in separate fuel tanks and then combined inside the rocket motor proper, where they ignited. But if the two fuels didn't mix in exactly the right proportions, or if any one of a number of other things went wrong, then an explosion was all but guaranteed. As Carolyn Henson once put it, "A rocket is a bomb. All it wants to do is blow up."

AMROC's rocket *couldn't blow up,* the reason being that it was to be made with a new type of propellant, one that combined both solid and liquid chemicals in such a form that would allow them to burn but not explode. "The propellant might fail to ignite," Bevin McKinney said. "The rocket might fall out of the sky like a brick. But the one thing it *can't do* is explode." That, at least, was the theory.

Not everyone, of course, was swept away by Gerry O'Neill's space colonies. Social scientists with strong attachments to earth seemed to be offended by the whole idea, and one critic, Jack Baird, a Dartmouth psychology professor, took particular exception to one of the more grandiose passages in Tom Heppenheimer's best-selling book, *Colonies in Space*.

Heppenheimer was an aerospace engineer who had first read the *Physics Today* piece as he was riding on a bus through Mexico. Not long afterward he too was attending the Princeton conferences, the NASA summer seminars, and all the other major events. Naturally he was a card-carrying L5 member, and wrote for the *L5 News,* including some short stories, which he called, quite appropriately, "Home on Lagrange." He also made the obligatory pilgrimage. "I went to Tucson to meet the Hensons, and found a goat farm and junked cars."

Heppenheimer's book pictured space colonies in rather glowing terms—not as utopias, exactly, but more like bucolic frontier communities that would be attractive to all sorts of special-interest groups. He wrote, for example, of American Indians going up into space to make new homes for themselves.

"We may see the return of the Cherokee or Arapaho nation, not necessarily with a revival of the culture of prairie, horse, and buffalo, but in the founding of self-governing communities which reflect the distinctly Arapaho or Cherokee customs and attitudes toward man and nature."

This was a bit much for Jack Baird, who had studied O'Neill's space colony proposals in some detail. Somehow, Baird could not quite picture a tribe of Arapaho winging its way through space in a large tin can.

"To the Native Americans, land is especially sacred," Baird said, "and today it is the particular land of their ancestors they would dearly love to recover and preserve for future generations. Circling the earth in a mammoth space station would hardly qualify as a promising spot from which to revive and pay homage to the traditions of their forebears. The native peoples who coexisted with the natural environment on equal terms for thousands of years before the abrupt arrival of the Europeans would be disturbed enough to see their nation confined to government reservations, never mind banished to outer space. For these reasons, an Indian reservation in space is probably not high on a list of Native American social goals."

Native Americans, indeed, had never been conspicuous advocates of spaceflight, much less space colonization. Neither were environmentalists, for that matter, this despite the fact that O'Neill had insisted from the beginning that protection of the environment would be one of the main benefits of space colonies. "If work is begun soon," he'd said in *Physics Today,* "nearly all our industrial activity could be moved away from earth's fragile biosphere within less than a century from now."

Apparently, though, none of that mattered to environmentally minded social scientists, who saw a large-scale migration into space as fostering a "disposable-planet mentality" as regards earth. Many environmentalists, in fact, were dismayed by the L5 Society's opposition to the Moon Treaty, a United Nations covenant that described outer space as "the common heritage of mankind," and contained clauses that, in the Hensons' eyes, would all but make space colonization impossible. Carolyn and Keith therefore took the L5 Society into a pitched battle against the treaty. As Keith described it in the *L5 News,* "On the Fourth of July 1979 the space colonists went to war with the United Nations of Earth. . . . The treaty makes no provisions for the civil rights of those who go into space. In fact, it authorizes warrantless searches. The treaty makes about as much sense as fish setting the conditions under which amphibians could colonize the land."

Somewhat to the surprise of their own membership, all the L5 lobbying had some effect—even the *New York Times* had admitted as much in its coverage of the story—for the United States Senate refused to ratify the Moon Treaty. From a legal standpoint, therefore, the pathway remained open for the great human exodus off the planet.

None of the major Space Rangers, neither O'Neill nor the Hensons, ever regarded their plans for the human race as especially hubristic undertakings. Space colonies only made good sense: they'd liberate people from ossified governments and gravitational chauvinisms; they'd confine pollution to hermetically sealed capsules far from earth; they'd provide humanity with new freedoms and "new neural perspectives"—whatever that meant. In any case, it would be an adventure. Was any of this really all that sinful?

It seemed not. There were far worse sins these Space Rangers could have committed. Soon enough, some were committing them.

3

Heads Will Roll

Jokingly, Saul Kent used to tell his mother, Dora, that when she died he'd drop her body down a hole in a park somewhere. She knew that he was only kidding about this. Still . . .

Dora Kent wanted to live forever, just like her son did, and both of them had been interested in cryonics, the practice of freezing the newly dead for later revival, ever since it got going, which was back in the 1960s. The two of them had been early members of the Cryonics Society of New York, and Saul had even participated in one of the first freezings ever done, in the summer of 1968.

That was in the prehistoric era of cryonics, when the freezings were done by funeral directors. The science had come a long way since then, and by 1987, when Dora Kent was near death in a nursing home, there was a whole new freezing technology in place, with its own inventory of patient stabilization protocols, blood cooling procedures, cryoprotectant ramp rates, and so on, all of it to prepare the body for its long trip to eternity. Many of these techniques had been pioneered by the Alcor Life Extension Foundation of Riverside, California, which was regarded by many cryonicists as the world's most advanced cryonics organization. In fact, one of the reasons that Saul Kent himself lived in Riverside was to be right next door to Alcor.

So when Dora Kent went into the nursing home, it was with

the understanding that she'd be frozen as soon as she died. Dora had been bedridden for the previous four years as a result of organic brain syndrome, a condition that had left her mentally incompetent and unable to take care of herself. In addition she suffered from osteoporosis and atherosclerosis. Then, in December of 1987, when she was eighty-three, she contracted a case of pneumonia.

The prognosis in view of all this was extremely poor, to say the least, so instead of ordering further treatment, Saul Kent withdrew his mother from the nursing home and had her transferred to the premises of the Alcor Foundation, which was just a few miles away.

The reason for the move was to save time: nobody knew whether a frozen corpse could ever be brought back to life again, but one thing everyone in the business agreed on was that the sooner after death the patient was frozen, the better his or her chances for eventual revival. Optimally, the freezing procedures would be started even *before* death, but under current law that would be homicide; the next-best thing would be to wait until the patient was pronounced legally dead, and then to begin with the freezing procedures. So when Dora Kent died at twenty-seven minutes past midnight on Friday, December 11, 1987, Saul Kent had his mother frozen.

Or at least part of her. The fact was that the maintenance costs for storing an entire frozen body for years into the future were quite high. How many years no one knew, but current thinking was in the neighborhood of one hundred or two hundred years, and to cover this plus the cost of the initial freezing, the sum of about $100,000 was required in advance. A separate consideration was the fact that Dora Kent's mortal body—with its hard arteries, soft bones, and diseased lungs—was not really worth saving anyway. The only part worth saving was the part that counted, where her personality and memory lay, which was of course her head. The theory was that if and when science ever got to the stage that it could bring a frozen person back to life, it could just as easily revive a frozen head and then attach it to a new body, perhaps one that had been cloned from the patient's own cells. Freezing and storage costs for a severed head—a "neurological suspension"— would be far less than for an entire body, only about $35,000. And so along about six o'clock on the morning of December 11,

Dora Kent's head was separated from her body, wrapped up in layers of protective fabric, and prepared for immersion into a tank of liquid nitrogen.

This was not an unusual procedure at Alcor—there were in fact six other frozen heads and one entire frozen body already on the premises—but nevertheless the suspension team on duty that night, which included Michael Darwin, president of Alcor, and Jerry Leaf, the staff surgeon, committed a minor technical blunder that would later turn out to have some major repercussions, not only for them personally, but for the practice of cryonics in general. The problem was that in their haste to get on with the freezing process, they failed to have the patient *pronounced* dead. That the patient was *in fact* dead (at least according to generally accepted medical standards) no one present had any doubt, for she had stopped breathing and had no heartbeat. Both Darwin and Leaf had verified this by means of cardiac monitor as well as by stethoscope, but neither of them was a licensed physician, so their clinical findings had no legal status. Nevertheless, because time was crucial, they immediately went ahead with the drug protocols and all the rest, and in a matter of hours Dora Kent's head was off and on its way to the "cephalarium vault," a special earthquake-resistant storage chamber that Alcor had developed for the better protection of its neuro-preservation patients (frozen heads).

The following Monday, officials at the Buena Park Chapel and Mortuary, with whom Saul Kent had contracted to cremate the rest of the body, tried to file a death certificate with the public health service in order to get a cremation permit. But the health service refused to issue one, because no physician had been in attendance at the time of death, and what was worse, the body in question was without a head, which was a highly unusual circumstance even for southern California. Two days later, people from the county coroner's office showed up at Alcor to examine the so-called nonsuspended remains. The coroner removed these from the premises, and later performed an autopsy on them.

Apparently, though, there would be no further difficulty. The autopsy showed that pneumonia was indeed the cause of death, and soon a deputy coroner signed a death certificate to this effect, listing atherosclerosis and organic brain syndrome as contributory factors. On December 23, with everyone satisfied, Dora Kent's body was cremated and the case was closed.

Or so everyone thought at the time. But at about noon on the very next day, December 24, the day before Christmas, an NBC camera crew from Los Angeles showed up at Saul Kent's home on the outskirts of Riverside. They wanted to know how he felt about the story in the paper.

"What story?"

"The story that you cut your mother's head off while she was still alive. And that now you're being charged with homicide."

*I*n March of 1987, about nine months before the Dora Kent affair, the L5 Society merged with the National Space Institute to become the National Space Society. The new society then had a total combined constituency of some sixteen thousand. The Hensons had resigned from L5 leadership years earlier, and in 1981 the two had gotten divorced. Carolyn resumed using her maiden name of Meinel, a name she retained even when, two years later, she married John Bosma, a defense analyst and Star Wars advocate.

After the divorce Keith Henson packed up and headed for Silicon Valley, home of the computer industry, the private launch vehicle business, and just about anything else that he found interesting. Before he left, some of his Tucson friends gave him a big send-off, sort of a going-away present. These were five or six of the old recreational explosives gang, and they drove him forty miles out in the desert to where they knew there was an abandoned, sixteen-hundred-foot-deep mine shaft. They poured two hundred pounds of liquid propane into the shaft, threw a lighted flare in after it, and got out of the way quick.

Even Keith Henson, who had seen many a recreational bomb go off in his day, was awestruck. "It was just absolutely gorgeous," he said later. "The flame must have gone up over a hundred feet in the air. It was the largest explosion I've personally had anything to do with."

By the time of the Dora Kent episode, Henson was living in San Jose, just south of San Francisco, and was married to Arel Lucas, who for a while had edited the *L5 News* back in Tucson. In the interim, Keith Henson's world view had undergone a major reorientation. On the one hand he'd become progressively disillusioned with the near-term prospects of space colonization, because increasingly nothing seemed to be happening, spacewise. NASA and the space shuttle were doing nothing more impressive than putting up a few unmanned satellites now and again, repairing one or two old ones in orbit, and taking a space walk or two—all of which was very much yesterday's news.

Private rocket companies such as OTRAG, Starstruck, AMROC, and others were springing up right and left, like mushrooms, and after a while the competition got so bad that Gary Hudson, founder and president of Pacific American Launch Services, Inc., and a man of considerable dry wit, wondered why there hadn't yet been any murders in the private launch vehicle business.

But despite all their frantic jockeying for venture capital, all their research and development, and all their spanking new prototype vehicles, not one of these rocket-for-profit outfits had put anything up into orbit, and few of them had managed to lift even so much as a single rocket off the ground. In view of all this it seemed to Keith Henson that the further you got toward the end of the twentieth century, the less and less likely it was that O'Neill space colonies would be orbiting up overhead within his own lifetime. He himself, at any rate, had just about given up hope of ever living in one.

Then, too, space colonies were never the whole of what Keith Henson really wanted anyway, they were just the best he thought he could get. You'd go up and live in orbit, grow food in space, and make a million dollars, all of which was fine as far as it went, but on the other hand living in a near-earth space colony was not all that different from living on earth. No sooner would you get up there, make your fortune, and start enjoying life than you'd be getting old, losing your marbles, and dying.

It was all rather tragic, actually. In his mind's eye Henson could see himself visiting the planets, going to the stars, crossing the Galaxy—in fact he'd been doing precisely that through science fiction ever since he could remember—and it was a shame that all of it should be closed off to him just because his genes had slated him for death in a few tens of years.

Naturally, an advanced thinker like Keith had long since heard about cryonics. In fact he'd heard about it from Saul Kent himself, who had been an L5 member and used to come down to Tucson to "help out," just like everyone else. Saul would come down and go through the tunnels . . . but since he also wanted to *live forever* he made a special point of getting back out again in a hurry. "I was very nervous about being in them," he remembered. "They were just earthen tunnels with nothing holding them up and you had the feeling they might collapse at any moment. I didn't stay down there too long."

Saul would tell Keith about how to live forever, but Keith, being a practical, no-nonsense type, thought the whole cryonics scheme was bogus. The freezing damage alone would be enough to turn your brain into something resembling a defrosted tomato, so what in God's name was the point of it all?

If anyone had the answer to this, it was Eric Drexler. Drexler, whom Keith and Carolyn had known since the first Princeton conference, was writing a book about a scenario he was just then coming up with for building a vast population of self-replicating, incredibly tiny robots, ones that were the size of individual molecules of matter. The virtue of their smallness was that the robots would be able to manipulate atoms one by one; they'd be able to grab hold of atoms just as if they were building blocks and place them into any physical arrangement that was allowed by nature. It would give you complete control over the structure of matter; you could turn coal into diamonds if you wanted to. In fact owning a fleet of these things would be like having your own mind-over-matter machine, a universal building engine: you could dial in a set of specifications, start the machine going, and pretty soon out would come whatever you wanted, all of it having been manufactured at the atomic level by Drexler's wee mechanical constructors.

Drexler had looked into cryonics, too, when he was still an MIT undergraduate, and his first reaction had been much like Keith Henson's, that it wouldn't work. For one thing, the person to be frozen would be dead to start with, which wasn't entirely promising. On top of that there'd be freezing damage (plus thawing damage, when revival time came), all of which combined to give

Drexler a somewhat jaundiced view of the enterprise. "Anybody who's taking this cryonics stuff seriously is probably crazy." That's what he thought to himself at the time.

But then he came up with his molecule-size robot scenario, and all of that skeptical negativity went by the wayside. To start with, a robot that could work with individual atoms could make repairs to damaged cells quite easily. Take frostbite, for example. Although frostbitten cells were obviously damaged, their gross structure was nevertheless well preserved by the cold. There ought to be enough intact material there for a fleet of tiny cell-repair machines to restore the cells to health and turn them back to normal functioning.

And if *that* was true, then why couldn't the robots equally well undo the freezing and thawing injuries caused by the process of cryonic suspension? It would take lots of programming to get the robots to do this, an extreme amount of it, software many orders of magnitude more sophisticated than anything ever before attempted or possibly even conceived of. All of it would be difficult, that was for sure, but the important point was, it wouldn't be impossible.

It was a long time, though, before Keith Henson could agree. Initially, the very idea of Living Forever made him distinctly uncomfortable. It *changed things around* too much. No sooner did he have a new life plan pretty well worked out for himself than he had to consider the prospect of maybe changing it once again. At least he'd have to if Drexler was right about his robots.

*T*wo weeks after the Riverside County coroner decided that Dora Kent's death might have been a homicide, some of his deputies showed up at Alcor with a search warrant. The medical examiner had found drugs in Dora Kent's body and now theorized that these drugs could have hastened or caused her death.

Drugs were routinely used during the course of a cryonic suspension: the procedure was to apply resuscitation measures as soon as the patient was pronounced clinically and legally dead, just as though the patient were going to be revived immediately. In fact, cryonicists used much the same methods that physicians employed in order to resuscitate patients who had temporarily stopped breathing, suffered cardiac arrest, or whatever else. The reason for this was to keep blood and oxygen circulating through the brain—which, after all, would be coming back to life someday. But of course cryonicists didn't want the suspendee coming back to life during the suspension process, so to ensure that the patient did not in fact return to life and consciousness, they injected the body with barbiturates and other drugs.

The crux of the matter in the Dora Kent case was whether the medical examiner could reliably distinguish between drugs that were administered immediately after death as opposed to a few moments before. This was at best highly speculative. Had a licensed physician been standing by when Dora Kent died, and had that physician pronounced her legally dead, the issue would probably never even have arisen. But since she had *not* been so pronounced, the coroner felt duty-bound to autopsy the whole body, including the head.

Such was the reasoning that led deputy coroners to show up at Alcor, search warrant in hand, ready to remove, among other things, all of Alcor's patient care records, the frozen patients themselves, and, specifically, Dora Kent's frozen head. But to the coroner's extreme surprise and embarrassment, the head was not to be found anywhere on the Alcor premises, nor would anyone there say where it was. From what the authorities could get out of Mike Darwin and Hugh Hixon—both of whom had been in on the Dora Kent suspension—they concluded that the head had been spirited away to parts unknown.

Darwin and Hixon were handcuffed. The authorities waited for some other Alcor people to return from lunch, who when they got back were also handcuffed. Finally the whole lot of them were taken to the Riverside County Jail, where they were fingerprinted, photographed ("mug shots"), and put into a holding area.

A few hours later all of them were released without being formally charged. The police made it clear, however, that they were still intent on getting the rest of Dora Kent. "We'll leave you alone," one of them told Mike Darwin, "if you just give us the head."

About a week later, the police decided that they were going to search the place again, only more thoroughly, so a Riverside County SWAT team arrived up at the Alcor lab along with some of the UCLA campus police. The claim this time was that Alcor was harboring thousands of dollars' worth of stolen UCLA medical equipment and supplies.

Many of the items the police took away did, admittedly, have UCLA identification tags on them, but Alcor people insisted that they'd purchased everything quite legally, from the university's own Surplus and Excess Property Department, and had records to prove it. Unfortunately, those records were now in the hands of the police, putting Alcor in a Catch-22. (Much later, the police would return the records and the property in question.)

Over the next thirty hours, the police stripped the place of everything that wasn't nailed to the floor or too heavy to lift (such as Alcor's scanning electron microscope). In all, the authorities hauled off eight computers, plus software and related paraphernalia—hard disks, backup tapes, and high-speed printers. They confiscated about five thousand dollars' worth of the drugs that were used in the suspension procedure, as well as Alcor's two German shepherds, although for what conceivable reason, no one there could guess.

They did not find Dora Kent's head. They did, however, find her hands.

Cryonics began in 1964 with the publication of a book called *The Prospect of Immortality,* by one Robert C. W. Ettinger. Ettinger was then a physics professor at a small college in Michigan, and had latched on to a notion that had been considered briefly by many others before him, although few ever took it as seriously as he did. Benjamin Franklin, for example, had written, way back in 1773, "I wish it were possible to invent a method of embalming drowned persons, in such a manner that they might be recalled to life at any period, however distant; for having a very ardent desire to see and observe the state of America a hundred years hence, I should prefer to an ordinary death, being immersed with a few friends in a cask of Madeira, until that time, then to be recalled to life by the solar warmth of my dear country! But . . . in all probability, we live in a century too little advanced, and too near the infancy of science, to see such an art brought in our time to its perfection."

Two hundred years later, though, science had most definitely advanced to the point where it was possible to regard storage of the dead body and its later revival as a workable proposition. At least Ettinger himself thought so, after having been primed for the idea by the potent combination of science fiction, mainstream science, and personal experience.

As a youngster Ettinger had been a fan of Hugo Gernsback's *Amazing Stories,* the pioneering science fiction magazine that Gernsback started in 1927. Many of the stories were vivid accounts of science and technology accomplishing all manner of wonders—interplanetary travel, metal men, suspended animation, and so on, but the story that the young Ettinger found most gripping was "The Jameson Satellite," by Neil R. Jones. It had been published in the July 1931 issue of *Amazing,* when he was only twelve, but Ettinger had remembered it ever since. It was about this Professor Jameson who'd stipulated in his will that after his death his body was to be fired into orbit, where it would be preserved indefinitely by the cold and vacuum of space. Millions of years later, with humanity long since extinct, a race of mechanical men discovered Jameson's frozen body, returned his brain to life, and transplanted it into a mechanical frame, after which Jameson lived forever.

Bob Ettinger liked that idea immediately. In fact, he wondered why we had to wait for the *aliens* to repair and revive frozen people and turn them into immortal beings. Why couldn't we do it ourselves? And why wait a million years? Science ought to be able to do it much sooner than that. Once he considered the subject for a while, though, he realized that he'd never heard about scientists actually doing any work on the problem or even so much as thinking about it. This was extremely puzzling.

War came, and Ettinger went over to Europe as a second lieutenant infantryman. He was wounded in Germany and returned to the United States, where he spent several years in an army hospital in Battle Creek, Michigan, recuperating and thinking about Matters of Life and Death. Here he learned that, finally, there *was* some work being done: he read about the French biologist Jean Rostand, who had frozen frog sperm and then revived it several days later.

That, at least, was a beginning. Rostand had also predicted that the day would come when the aged and infirm would be put on ice to await help in the future. That was more like it; at least *someone* was thinking the right thoughts.

Also while in the hospital, Ettinger wrote a science fiction story of his own, "The Penultimate Trump." This was published in the March 1948 issue of *Startling Stories* ("A Thrilling Publication"). In its own way, it was an extraordinary document. The story was about a millionaire, H. D. Haworth, a foul man who'd come by much of his money dishonestly. He'd used his millions to prolong his life as much as possible, but then finally, at the age of ninety-two, he was on the verge of death. A scientist by the name of Stevens, meanwhile, had been experimenting with suspended animation techniques and had finally gotten somewhere. His method involved "premilimary partial dehydration of the living tissue through starvation, followed by freezing."

So Haworth had himself frozen and placed in a vault. "He was the first of men," the narrative said, "to die a qualified death."

Three hundred years later Haworth was revived and returned to a state of full-blooded youth. "*Well, this is it,*" he thought to himself when he woke up, "*it actually panned out. I did it, I did it!*"

The revival and rejuvenation procedures had taken a long time. "Two years we've been working on you," one of the doctors told him.

Haworth wanted to get on with his life immediately, but first there were one or two matters to attend to. During his frozen three hundred years, scientists had perfected a method of reading people's thoughts. They could do this while you were still frozen, with, "in simple terms, a hypno-biophysical technique for reaching and interpreting buried memories."

This was a bit ominous. "Your thoughts and experiences are on file," the authorities told him, "and the newsworthy ones have been published."

Also a surprise! But there was even worse to come, because this thought-reading stuff had been turned into a behavior control system. ("What it really did," they told him, "was take most of the irresponsibility out of people's behavior.")

The system worked by meting out punishments for past infractions (as read out of your buried memories). All this was done with perfect mathematical exactitude. "*Suffering* has been classified, qualitatively and quantitatively," they said. "If he's passed a certain allowable maximum in wrongdoing, a person must go to the penal colony and experience himself all the suffering he has caused, qualitatively and quantitatively as closely as possible."

And so for the evils done in his earlier life, H. D. Haworth was sent away to the penal colony, located on a planet that used to be called Mars.

"Now they call it Hell."

*B*ob Ettinger attended the University of Michigan and then Wayne State University, where he got bachelor's and master's degrees in physics, plus a separate master's degree in mathematics. Later he taught physics and math at Wayne State and then at Highland Park Community College, in Michigan. It was always in the back of his mind to make out a scientific case for cryonics, but he never did anything about it until 1960, when he had his midlife crisis.

"By 1960," Ettinger said, "it was clear that the Abomination was indeed gaining on me, so I wrote up the idea in a few pages." He sent this around to a few hundred people whose names he'd gotten from *Who's Who,* but nobody paid any attention. It was as if they didn't care whether they lived or died.

"People had to be *coaxed* into realizing that dying is (usually) a gradual and reversible process," he said. "For that matter, a great many people have to be *coaxed* into admitting that life is better than death, healthy is better than sick, smart is better than stupid, and immortality might be worth the trouble!"

So he wrote a whole book, *The Prospect of Immortality,* to *coax* people into admitting these things. Ettinger published the book on his own in 1962. Two years later a revised version was bought out by Doubleday. *The Prospect of Immortality* ultimately went through nine editions in four languages and became the bible of the cryonics movement.

The book gave an item-by-item account of all the scientific evidence Ettinger could find that, at least in principle, the frozen dead could be brought back to life. For example, many species of insects and lower organisms were known to freeze solid during the winter and then revive again automatically in the spring; some of them secreted their own natural antifreeze, glycerol.

Then there were the experiments done by Audrey Smith, in England, on golden hamsters. She'd cool them to the point where about half of their body water turned to ice and then warm them back up again, and many of them survived without any impairment whatsoever. Similar experiments had been done by others, on rats and on dogs; some of the animals had survived, others hadn't. There was lots of evidence like this, none of it conclusive proof that frozen mammals would be revivable, but still and all, he thought, the record seemed far more positive than negative.

And so in his book Ettinger proposed actually taking the gamble. If freezing didn't work, *so what?* What did a person have to *lose?* You'd *already be dead,* which was its own worst punishment.

Ettinger himself, however, thought that advanced science would almost certainly be able to bring you back. Everything hinged on the proper repairs being made to damaged structures beforehand, but there was no reason to think that the necessary techniques would not be developed: "Surgeon machines, working twenty-four hours a day for decades or even centuries," he wrote, "will tenderly restore the frozen brains, cell by cell, or even molecule by molecule." The frozen person will be resurrected, "alive and in much better health than just before he died." Youth will be restored: "You and I, as resuscitees, may awaken still old, but before long we will gambol with the spring lambs—not to mention the young chicks, our wives."

Ettinger became an overnight media star. He appeared on radio and television; magazine articles and newspaper stories were written about him. The idea of cryonics suddenly gained a vast audience, some of whom would soon take it upon themselves to put his ideas into practice.

When Ettinger appeared on "The Long John Nebel Show," an all-night radio talk program with audience call-ins, who should be listening in on the airwaves but Saul Kent. A college student at the time, Kent thought that the freezing idea was absolutely correct. Indeed, *how could anyone be against it?* "It hit me instantly," he remembered. "It was like a lock waiting for a key to be put in."

Kent listened to the whole program and the next day went out and bought a copy of Ettinger's book, took it to the beach—Saul Kent spent many a day there when the weather was good—and read it from cover to cover. That was in June of 1964. A year later he and some other like-minded souls founded the Cryonic Society of New York, and three years after that the group did its first suspension, freezing one Steven Mandell, who never even got to old age. He died at the age of twenty-four.

Mandell, however, was not the first man ever frozen for the purpose of later resurrection. That distinction belongs to James H. Bedford, a retired psychology professor, who was frozen in January of 1967 by Robert F. Nelson and friends, in California.

Nelson, who in a checkered career had started out as a prize-fighter, was part owner of an electronics shop in the San Fernando Valley when he read about Ettinger's book in a newspaper. It was another case of a lock waiting for a key to be put in, because Nelson too went out the next morning, bought a copy of the book, and polished it off in a single sitting.

As crazy as it might have sounded to some, Nelson decided that this was something that had to be tried. "The idea was at once preposterous and so entirely logical," he recalled. It was so entirely logical that three years later Nelson was doing his first freezing.

The thing was a nightmare from beginning to end. First, the patient died sooner than expected.

"*The patient died?*" Nelson said when he heard.

And at that important juncture the physician who was going to do the actual freezing, a Dr. Mario Satini, started having some last-minute doubts.

"I wanted to be at the patient's side when he died," Satini told Nelson. "I don't know. I should talk to some people first. I don't know if I should do it."

And finally, after the deed had been done, and as they were transporting the body from one place to another (no one wanted to keep it for very long), it became clear that they were . . . *running low on dry ice!*

Nelson decided to refill Bedford's coffin in transit. He arranged to meet a friend of his at a public park, where they'd open the casket and pour in a fresh supply.

"There we were in broad daylight," Nelson recalled, "in full view of children swinging as their mothers looked on, a group of young boys playing ball, a city employee mowing the lawn, and a young couple strolling by holding hands—lifting the lid of what could be nothing but a coffin, clouds of smoke arising from it, and not one person even gave us a strange look."

After the publication of *The Prospect of Immortality*, the scientific evidence for cryonics continued to mount. There were the frozen brain experiments, for example.

In the mid-1960s, three Japanese professors, Isamu Suda, K. Kito, and C. Adachi, of the Kobe Medical College, froze and thawed out a series of cat brains. The brains were drained of blood, filled with glycerol, and frozen for periods ranging from overnight to six months. Then they were thawed out, and the glycerol was removed and replaced by fresh, warm cat blood. The researchers fixed electrodes to the gray matter and hooked them up to a tracing apparatus much like an EEG (brain wave) recorder.

Given the conventional wisdom that brain cells couldn't survive for more than a few minutes without oxygen, what happened after that was an utter revelation. In the finest schlock-horror-movie tradition *the tracing needles started to move!* They scratched back and forth, up and down, describing *waves,* just as if they were picking up new electrical activity within the brain.

And indeed there was no other explanation. It was almost impossible, well-nigh incredible, in fact, but the evidence was right there in front of them as the needles continued to write their eerie message: this six-month-frozen, barely defrosted cat brain was putting out brain waves, and furthermore of a kind not substantially different from those it had generated before it was frozen. More strangely still, the brain was doing this of its own accord, spontaneously, without any electrical shocks having been applied to it, without benefit of any artificial stimulus whatsoever. It had been thawed out, refilled with blood, and then . . . it simply *started running again.*

Suda and his colleagues published their results in the eminently respectable British science journal, *Nature.*

"At this stage we wish to conclude," they said, "that brain cells are not exceptionally vulnerable to lack of oxygen. It appears that even nerve cells of the brain can survive and be revived after long-term storage under special circumstances."

Nor, indeed, was this an isolated case. Similar studies were done later, by others, on monkey and dog brains, with essentially the same results. And there were experiments on larger mammals, some of which became media stars in their own right, like Miles the beagle.

Miles wasn't frozen, he was only *cooled*, to within a few degrees of freezing, but all of his blood had been drained out and replaced by a blood substitute. He was held at about 34°F for about a half hour, completely bloodless. Then he was warmed back up to room temperature and his original blood was restored. Eventually he got up and walked away, and then he was again as normal a dog as you could find anywhere.

All of this was done by Paul Segall, a researcher at the University of California at Berkeley, and a director of the cryonics firm Trans Time, Inc., of Oakland. Afterward, Miles was featured in *People* magazine, and, together with Segall, appeared on "Phil Donahue." Everyone loved him, Miles the wonder dog.

Then there was the example of the frozen frogs to think about. It was discovered that several kinds of frogs in the northern United States and Canada freeze during the winter and come back to life again the following spring. This too, according to cryonicists, is proof that "frozen" does not mean "dead." *Enemies* of cryonics, such as James Southerd of the University of Minnesota, had an answer to this. "Frogs are made to be frozen," Southerd claimed. "It's the way they are. People, unfortunately, are not made to be frozen."

The fact was, though, that by the late 1980s, some people had been frozen, defrosted, and had come back to life. As in the case of John Brooks.

Brooks was conceived (at room temperature) in a laboratory dish at Monash University, in Melbourne, Australia. So far, this was an ordinary case of in vitro fertilization, but the difference was that after the zygote went through a few cell divisions it was placed into a liquid nitrogen bath at −196°C for a period of two months.

(Cryonics patients were being stored in the same liquid nitrogen and at the identical temperature.) Then, at the end of the two months, the embryo was transferred to the womb of Margaret Brooks, John's mother, where it implanted successfully. Nine months after the implantation—but *eleven* months after his conception in a lab dish—John Brooks was born, weighing in at eight pounds, two ounces.

By 1987 frozen human embryos had become common enough that the Vatican issued a statement denouncing the practice: "The freezing of embryos, even when carried out in order to preserve the life of an embryo—cryopreservation—constitutes an offense against the respect due to human beings."

Two years after the Vatican statement, though, people were going to court over who had ownership rights over the frozen zygotes. One court case, in Maryville, Tennessee, was brought by a couple who did the in vitro part while they were still married, had the eggs frozen, and later got divorced. The wife claimed rights to the zygotes, but the husband, who didn't want to be held responsible for child support in case they were implanted, wanted them destroyed. (In September of 1989 the court ruled in favor of the mother, Mary Sue Davis, who was awarded custody of the seven embryos. The father, Junior Lewis Davis, planned to appeal.)

By this stage, human embryos had stayed frozen for over two years—twenty-eight months was the record—and then successfully implanted. No one knew what the outside limits would be, if indeed there were any. To cryonicists, the message seemed clear that "frozen" didn't necessarily mean "dead," even if you were talking about human beings and not just frogs.

Of course, that was if you froze a *live* person to start with, which was not what the cryonicists did. "The problem is, they always start with a dead person," said Jim Southerd. "It might make more sense if he were alive, but he's not, he's dead."

Cryonicists of course had to agree that your best chance is to be frozen prior to death, maybe while you're even in the pink of health. Others might see this as a tragedy, but not a dyed-in-the-wool cryonicist. Leo Szilard, the physicist, once wrote a science fiction story about freezing people, "The Mark Gable Foundation." In it, the narrator gets frozen while he's alive.

"I spent my last evening in the twentieth century at a small farewell party given to me by friends. There were about six of us, all old friends, but somehow we did not understand each other very well on this occasion. Most of them seemed to have the feeling that they were sort of attending my funeral, since they would not see me again alive; whereas to me it seemed that it was I who was attending their funeral, since none of them would be alive when I woke up."

*F*inally some evidence about the success of cryonics emerged from among the ranks of the frozen themselves. Not that they were *revived*, they were only *defrosted*. Or at least parts of them were.

In November of 1983, for the first time ever, a cryonics firm conducted autopsies on the defrosted mortal remains of two Trans Time patients who had been converted to neuro. The parties in question were a married couple who had wanted to be frozen after death. They hadn't had the required lump sum in advance, but Trans Time didn't want to turn them away, so it accepted them on a contingency, pay-as-you-go basis. Monthly maintenance costs would be covered by their loving son.

When the son's parents died, Trans Time suspended them as whole-body patients, which was their desire, and for a number of years everything went along exactly as planned. But then the son himself died, in an automobile accident, after which the monthly payments ceased. Trans Time kept the parents frozen for a while, but it was clear that sooner or later something had to be done. It was a private company, operating without government support—indeed, often in the teeth of government opposition—and could not afford an extended period of unpaid maintenance, especially when for fifteen out of its sixteen years of existence the company had run at a loss.

But then the Alcor Life Extension Foundation came to the rescue. It would take care of the frozen parents, essentially on a charity basis, but only on the condition that they could be "converted" first, which is to say, converted from whole-body to neuro, the latter being far less expensive than the former. "The same capsule that you put a whole body in," Saul Kent once explained, "you can probably put twenty heads in."

This, of course, meant that the heads had to come off while the patients were still frozen. Not that this was much of a problem. As Alcor's illustrated report on the case explains, "a rapid conversion to neuropreservation was done using a high-speed electric chain saw."

This was now a golden opportunity to see how frozen bodies actually fared over their years in storage—nine years for the husband, five for the wife. So the Alcor men thawed out and autopsied the newly decapitated bodies.

There was both bad and good news. "The most unexpected finding as a result of these autopsies," says the report, "is the discovery of serious fracturing in all of the suspension patients."

There were fractures in the outer skin, in the subcutaneous fat, in the blood vessels next to the heart, in the arteries and veins. The right lung of one patient was cracked almost in half, as was the liver, and there were open wounds on the hands and right wrist. This was not encouraging, but it was all too easy to lose one's perspective. The fact of the matter was that the injuries suffered by these frozen corpses were no worse than what's seen in hospital shock-trauma units every day of the week—broken (if not absent) arms and legs, and so on—but many of these people end up recovering. The fact that the frozen corpses were not in pristine shape was not by itself any cause for alarm.

The good news was that much of the bodies survived perfectly intact. The palms of the hands, the soles of the feet, and other structures were all in fine shape. As for the brains, they remained in suspension and were not examined.

To Alcor, the whole thing was a learning experience. The initial suspensions had not been perfect, but all things considered, the patients came through the whole process about as well as anyone could expect.

Supposing that eventual revival would be *possible,* the question remained whether it would be *desirable.* This had been considered time and again by science fiction writers, many of whom, surprisingly, answered in the negative.

In 1967, for example, Clifford Simak published a novel focused on cryonics, *Why Call Them Back from Heaven?* It was A.D. 2148, and billions of frozen people were packed into a New York City skyscraper known as "the Forever Center." The chances of eventual resurrection were so good that immortality was a virtual certainty except for the most hardened criminals, whose punishment was being allowed to die a natural death . . . *and never being revived.*

According to the narrative, cryonics "had started less than two centuries before—in 1964, by a man named Ettinger. Why, asked Ettinger, did man have to die?"

Now, in 2148, you didn't, but this was not necessarily a blessing. One of the drawbacks was that people took no chances during their so-called first life cycle: "No chances—no chances of any kind that would threaten human life. No more daredeviltry, no more mountain climbing, no more air travel, except for the almost fool-proof helicopter used in rescue work, no more auto racing, no more of the savage contact sports."

And so in the novel, people went through their first lives with a policy of exaggerated protectiveness, afraid of losing their only chance at eternity—and one of exaggerated economy as well, since they put all their money away for their next life. In fact it was as if their entire natural life span had become mere preparation, an overture to the day when they'd really be able to get on with their lives.

And then, when you finally *did* get thawed out, would immortality turn out to be all it had been cracked up to be? Simak, for one, didn't think so. There will be so many people alive, he said, that "the earth will be just one big building." In fact, the population problem would become so acute that humanity would be forced to invent time travel to deal with all the excess people. "Once they get time travel, they'll send people back a million years to colonize the land, and when the land back there is filled, they'll send them back two million."

Whole new psychological counseling industries would spring up in the future, the narrative said, their only goal being to ease the reawakened into the new century. There would be grad schools, Schools of Counseling, with stiff entrance exams and rigorous training programs, all of them designed to help you out on Revival

Day and whenever you might need help thereafter. Traveling into the future, Simak intimated, had to be regarded as an adventure, a cosmic experience like Columbus landing in the New World or an astronaut stepping out on another planet.

Indeed, this was one of the greatest unknowns in cryonics: Will you like the world you wake up into? Even some of the world's most forward-looking thinkers have found this difficult to answer. When Freeman Dyson's father died at the age of eighty-two, Dyson wondered whether he ought to have the body frozen. In the end, he decided not to go through with it. "He died gracefully," Dyson said, "and I decided that I would not be doing him a favor by bringing him back in a hundred years."

As for Dyson himself, he wasn't much interested in being frozen and resurrected either. "I have no wish to go through all the bother of dying twice," he said.

Robert Heinlein, apparently, felt the same way. His novel *The Door into Summer* dealt with cryonic suspension, which in the novel was called "Cold Sleep." The hero of the novel didn't even *want* to be frozen; he was *tricked* into it. "Sure I wanted to see the year 2000," he says at one point, "but just by sitting tight I would see it . . . when I was sixty, and still young enough, probably, to whistle at the girls. No hurry. Jumping to the next century in one long nap wouldn't be satisfactory to a normal man anyhow—about like seeing the end of a movie without having seen what goes before."

When, after having been frozen against his will, he's revived some years later, things are no better than he expected in the first place. The future "is not, repeat *not,* paved with gold," he says. Traveling to the future, like traveling to another country, made for homesickness, "that terrible homesickness which is the great hazard of the Sleep."

The cryonicists themselves, of course, had never been much impressed by any such objections. To them, it was simply a case of people being victimized by their culturally inbred *Deathoid Prejudices,* often to the point that many of them were convinced that not only was death inevitable, but that it was even *a good thing.*

Thus people objected to immortality with what were, to the cryonicists, endless beside-the-point questions, such as: *Won't eternal life be boring?*—to which the cryonicists had whole rafts of canned answers, most of which were entirely irrefutable by any known stratagem.

One: "Ordinary life is sometimes boring. So what?"

Two: "Eternal life will be as boring or as exciting as you make it."

Three: "Is being dead more exciting?"

Four: "If eternal life becomes boring, you will have the option of ending it at any time."

The critics asked, *If everyone lives forever, won't that contribute to the overpopulation problem?* and the cryonicists (in this case Art Quaife, president of Trans Time, Inc.) answered: "That's such a dumb point. Why don't you commit suicide to help the population problem? Yes, in the long run, cryonicists will contribute to the population. But we are not limited to planet earth."

Sometimes the critics even spoke of death as being *natural,* as if this was *what Mother Nature had in mind* for you all along, and that when your time came it was best to go quietly. To a practicing cryonicist, that kind of thinking was the ultimate horror. "Polio is also natural," said Jack Zinn, of the American Cryonics Society. "You don't see people condemning braces, crutches, or artificial hearts. What's natural is for man to be alive and happy."

No, the cryonicists were not exactly reduced to Silly Putty by the culture's ingrained *Deathoid Prejudices*—even including those of Clifford Simak or Robert Heinlein, whom they admired for at least *considering* the question of eternal life on earth. They were more impressed by the things those authors *got right* in their stories, such as the time a Heinlein character says, "Make sure that you take your Sleep in the Riverside Sanctuary in Riverside."

Heinlein wrote those words in 1957, thirty years before Alcor moved there, and one cryonicist was affected enough by this coincidence that he actually wrote to Alcor about it. Heinlein's prophecy was "so remarkable," he wrote, "that one must wonder whether Alcor deliberately moved to Riverside to take advantage of it. Did you?"

"No," replied Alcor president Mike Darwin. "Actually we moved to Riverside to take advantage of the sympathetic coroner."

*T*he coroner in Riverside County at the time of the Dora Kent affair was Raymond Carrillo. Carrillo had achieved a certain measure of fame earlier that same year, 1987, during the Liberace case. When Liberace died that February the announcement given out by the pianist's representatives claimed that death was due to congestive heart failure brought on by "subacute encephalopathy," the latter being the medical term for degenerative brain disease. But because no physician was in attendance at the time of death, Carrillo requested an autopsy. One of the possible causes of subacute encephalopathy, Carrillo knew, was AIDS.

Liberace's body had been sent to Forest Lawn, where it had already been embalmed, but the coroner got some tissue sections and from them was able to determine the true cause of death. He made his announcement a couple of days later, from the steps of the Riverside County administrative building.

"Mr. Liberace did not die of cardiac arrest due to cardiac failure brought on by subacute encephalopathy," Raymond Carrillo said. "He died of cytomegalovirus pneumonia due to or as a consequence of human immunodeficiency virus disease. . . . In layman's terms, Mr. Liberace died of an opportunistic disease caused by acquired immune deficiency syndrome."

When, ten months later, Carrillo had another unattended death on his hands, one which furthermore took place in a cryonics lab, an autopsy was clearly called for, and one could hardly blame the man for wanting to do the full procedure, head and all. By the end of 1988, however, it was becoming increasingly clear that some very strange things, deathwise, were going on up in the coroner's office.

On July 17 of that year a woman's body was found in the desert near Palm Springs, in Riverside County. It looked like a homicide to police, so the body was brought to the coroner's office, where an autopsy was scheduled for three days later. By the date of the autopsy, though, the body in question had vanished. The fact was, it had been released by mistake to a mortuary in the area, which cremated the remains before the slip-up was discovered. The mortuary, which had been expecting a man's body, failed to open the body bag and check the sex or identity of the person inside, but went ahead and cremated it anyway. So now the police had a probable homicide case on their hands, but no body to show for it, and no coroner's report as to the cause of death.

The coroner blamed it all on the funeral home. "They should have looked inside the bag," Raymond Carrillo said.

The funeral home, for its part, blamed the coroner. "We feel the responsibility is with the county coroner's office," said Dennis Butler, owner of the Rubidoux Mortuary.

After an investigation, Carrillo officially fixed responsibility for the mishap on Supervising Deputy Coroner Dan Cupido, whom he suspended for three days in punishment. Shortly thereafter, though, Cupido was in trouble again.

On March 2 of the same year, Cupido had acted as witness to the signing of the last will and testament of Jack Cook, himself a former Riverside County deputy coroner. Not only was he a witness, Cupido was also named in the will as *executor*, as well as sole *beneficiary* of an estate valued at approximately a half million dollars. Unfortunately, California law prohibits a beneficiary from being a witness to a last will and testament. In September, about five months after Jack Cook passed away, six of Cook's relatives sued Cupido for the estate plus punitive damages, alleging that he used undue influence with Cook to get himself named as sole beneficiary.

All these troubles made for lots of good times and black humor among Alcor officials, who began to joke about the coroners being Keystone Kops, Katzenjammer Kids, and so forth. But the events so far were nothing compared to what came down in the month of October. That's when it emerged that Brad and Didi Birdsall, husband-and-wife deputy coroners of Riverside County, had been doing moonlight autopsy work at their home, in the garage and backyard, using a picnic table as their lab bench.

This unpleasantness came to light the day after the Birdsalls sold their house to Mark and Gail McClure. When the McClures moved in, they found a note from the Birdsalls handwritten on a paper towel: "We still occupy half the garage, the covered patio, part of the brick patio (under the olive tree), and still have our tools in the shed. Call us if there any problems."

Well, there were some problems, all right: body parts were turning up everywhere. They were in little plastic bags that had a squishy feel to them and smelled of formaldehyde. They were labeled with people's names, dates, and such fear-inspiring words as "heart" and "stomach and contents."

The McClures called the cops, and later the same day Rick Bogan, another of Riverside County's seemingly endless supply of deputy coroners, came by in a car and took the offending materials back to the office. The body parts filled up the car's trunk and the backseat, as well as part of the front seat. The McClures had concluded, at that point, that their problems were over, but *No!* They went back and found more problems in the garage, in various boxes, bags, and buckets. By the time it was finally over that very bizarre Sunday, twenty-five boxes, bags, and buckets of human body parts had been taken from the McClures' house.

Raymond Carrillo, aided and abetted by Dan Cupido, now launched an official inquiry. In response to a reporter's question, Cupido said he didn't know whether a residential picnic table was a sanitary enough place for tissue sectioning work, although he admitted that, had he known about it at the time, "I probably would have objected to it."

Brad and Didi Birdsall, meanwhile, were referred to in the newspapers as "Riverside County's Fun Couple." ("They've given 'part-time job' and 'piecework' entirely new meanings," one columnist said, adding that the Birdsalls' marriage vow had been "Til death we do parts.")

People at Alcor were beside themselves. This was unbelievable. In a news conference earlier that year, Ray Carrillo had answered the charge, made in a newspaper story, that he was going to locate Dora Kent's head, "melt" it, and "chop it up."

"We don't chop up heads, we do autopsies," Carrillo said at the time.

But Alcor members had to wonder. God only knew what could have happened to Dora Kent's head had they turned it over to Tonto and Kemo Sabe up there in the county coroner's office.

A few weeks into the Dora Kent debacle, Mike Darwin resigned as president of Alcor. He was thirty-two years old at the time and had been president for six years. He was, in fact, the boy wonder of the cryonics movement.

Like lots of other kids, Mike Darwin used go around his neighborhood in the summertime collecting honeybees. He'd catch them in jars, store them overnight, and the next morning they'd be dead. But even a twelve-year-old knew about "suspended animation," so one time Mike put the jar in the refrigerator overnight and the next morning the bees were still alive. They moved more slowly than beforehand, but nevertheless they lived. He discovered, in fact, that you could store bees for two or three nights in the refrigerator, and they would come out of the icebox apparently none the worse for wear.

The next logical step was to put them in the freezer, but unfortunately that killed them. But then he tried the same thing with turtles, the little baby turtles that you could get in the dime store for fifty cents or so. He'd put these in the freezer for a half hour, by which time their limbs were so stiff that you couldn't move them at all, and even the neck and head were almost frozen solid. But if you dropped them in a glass of warm water, pretty soon they'd be swimming around again, good as new.

The question was, were they *really* being frozen solid, all the way through, or were they just suffering some sort of advanced hypothermia? Mike just *had* to know the answer to that, so when the turtles froze to the point that they showed no signs of consciousness, he took a scalpel, cut a hole in the bottom half of the shell, and glued a glass coverslip over the opening.

"That allowed me to actually see into them, and to watch their hearts," Darwin said. "And in fact I discovered that I was *not* freezing them solidly, I was only freezing about fifty percent of the water. But in their limbs and head—in the brain—I was probably freezing more than that."

Then, on his own, Mike began injecting some of the turtles with glycerol, to see if that would keep them alive longer in the freezer, but it didn't: thirty-five minutes in the freezer was still the outside limit. Exceed that, and the poor creature was gone forever. "On the other hand, when I dissected the turtles afterwards, the internal organs always looked great. That was because of the glycerol."

All this as a twelve-year-old. It was also about this time that Mike saw his first human corpse, albeit in rather unfortunate circumstances. He'd gone over to visit a relative and noticed that newspapers and bottles of fresh milk had piled up out in front. He broke open the screen door and found his aunt's five-day-old body decaying in the summer heat.

About a year later, a friend of his who knew about the turtle-freezing experiments brought him a newspaper clipping telling about the suspension of James Bedford, in California. Mike Darwin thought this was incredibly stupid; his own experiments told him so.

"I thought it was a crazy idea. Freezing does a lot of damage, I knew that. I had done microscopy in my turtles' tissues, from the legs and internal organs, and ice was disruptive. To me you had to have suspended animation; there was no point in freezing someone once they're dead."

But a couple of months later, Mike's father, who was a police officer, showed him an article in *True Man's Magazine*. It was about Bob Ettinger, about how he claimed there was all kinds of scientific backing for cryonics. Mike was so intrigued by this that he wrote letters to Ettinger as well as to all the other people mentioned in the article, one of them being Saul Kent, who by then was involved in the Cryonics Society of New York.

"Saul sent me back a huge packet of information," Darwin recalled. "And right then and there, at the age of thirteen, I decided that this was what I wanted to do with the rest of my life."

By the time he was seventeen, Mike Darwin had constructed a human suspension apparatus in his parents' house. Basically it was only a dry-ice box and a unit for flushing out people's blood and replacing it with glycerol—apparatus of about the same order of sophistication as what had been used in the very first freezings. In fact, when he later visited the Cryonics Society of New York,

Darwin was surprised to discover that his own equipment was better than theirs.

"The place was a *dump!*" he recalled. "It was a *shambles!* The patients were being well cared for, but otherwise it was just a nightmare. The people running it were unprofessional: one was a beachcomber and the other was an attorney who worked nights in a record factory rather than practicing law. And this scared *the hell* out of me! This was all that was standing between me and *everlasting oblivion!*"

Nonetheless, Darwin was in there pitching whenever the New York cryonicists did a suspension. Ten years later, when Darwin was the president of Alcor, even Dan Cupido, of the famous Riverside County Coroner's Office, had to admire the place. "They're better stocked than some medical facilities," he allowed.

What they were doing with all that equipment, Mike Darwin will tell you, was preserving information. "The essence of an individual is information," he said. "Cryonics is about storing a person's molecular framework, locking all that up and storing it indefinitely."

The cryonicist view is that life is *not* a process of constant activity, it's *not* a process that, once interrupted, is gone forever. Life is present, at least potentially, whenever there's enough information left to get the whole process started again; and freezing, they claim, preserves that information. How else to understand why the frozen cat brain starts running again of its own accord? How else to understand why the frozen embryos spontaneously come back to life when placed in the womb? What freezing preserves is a structure that, once unfrozen and let go, can blossom again of its own power. How ironic, then, the cryonicists think, that some people see their enterprise as "gory," as in the case of Dora Kent's severed hands.

"A butcher shop is gory," Mike Darwin said. "People are actually eating dead animals, and they don't even need to do that." (Mike is a vegetarian.) "A better example of *gory* is, you take a guy where everything is there—his mind, his personality, his memory, everything's intact, only he's 'dead'—and then you haul him off to a place where you cut him open and fill him up with formaldehyde, causing enormous structural disruption, and then you wrap him up like so much garbage and bury him, and he molds and rots, and microorganisms eat him up, and worms and insects pour into him, and he disappears forever. That's gory.

"Or you take somebody and you put him in a crematorium and you listen, and after about fifteen or twenty minutes at seventeen hundred degrees Fahrenheit, you hear this '*Pop!*'—that's his head blowing up because the water's boiling inside his head at a rate so fast that it can't escape, and so the head explodes. That's gory."

As for Dora Kent's hands, there was, according to Alcorians, a perfectly good reason for removing them. At age eighty-three, Dora Kent was the oldest patient Alcor had ever placed into suspension. Her arteries were calcified and clogged with plaque, and because of this the glycerol antifreeze was going through her body unevenly. There wasn't enough time to monitor this fully during the suspension itself, but there would be plenty of time afterward, as long as parts of the body were put aside and preserved. So for the sake of studying their glycerol concentration levels later on, the surgeons removed Dora Kent's hands along with her head.

On the night of May 8, 1988, Mike Darwin, who wears a telephone pager, got a message that an Alcor client had "deanimated" at his home in southern Florida. Darwin was attending a wedding reception in Pasadena at the time, but he left there at once and headed back to Riverside, to help prepare the lab for the patient's arrival.

The man, whose name was Bob, was packed in ice, flown to Los Angeles, and brought to Alcor in a van. By the time he arrived, the suspension team was assembled and ready to go to work. Besides Mike Darwin, the team included Jerry Leaf (chief surgeon), Brenda Peters (assistant surgeon), Carlos Mondragon (the new Alcor president), Mike Perry, Hugh Hixon, Arthur McCombs, Saul Kent, and Keith Henson.

After years of talking with Eric Drexler about his submicroscopic robots, and reviewing several drafts of Drexler's book, *Engines of Creation,* Henson had finally undergone the radical change of world view that he'd been afraid of making earlier. *Engines of Creation,*

Henson now thought, showed that cryonics could be a working proposition. In fact, Drexler's book showed in great detail how his little mechanical marvels could remake the universe from the ground up. If they could cure frostbite, they could revive a cryonics patient. Henson finally decided that he might as well take advantage of what seemed to be an ever surer bet. *Besides,* he thought to himself, *it would be really stupid to be one of the last people to die.*

So Henson signed up with Alcor for a neuro (a "head job"), as did his new wife, Arel Lucas. They also signed up their daughter, Amber, who at age two was the world's youngest signed-up cryonic client. After a while they'd even convinced many of their friends, including Timothy Leary, who was then living in Beverly Hills, about an hour away from Alcor, to sign up for head jobs. In fact, Keith was by now so excited about cryonics that he wanted to see a suspension firsthand, so when he got the call from Arthur McCombs at Alcor, he flew down to Riverside and was in full surgical dress by the time the patient was wheeled into the operating room.

Like most other cryonics patients, Bob was not rich, famous, or particularly unusual in any way. He'd been a TV repairman and family man, and without any doubt the most distinctive thing about him was that he wanted to come back in the future and live forever.

Bob was not going for neuro. Probably believing that there's more information in the body than resides in the brain alone (a controversial question even among cryonicists), Bob had elected to go whole-body. This would cause a bit of a problem. The patient had undergone two coronary bypass operations, and the area around his heart was, as the Alcor surgical team discovered when they got there, a mess. It took them three hours of dissection just to reach and identify all the major structures involved.

This was Keith Henson's first suspension, so he had the dirty-scrub duties, mopping up ice water, vacuuming up bone chips, dumping fluids down the drain ("Real blood-up-to-the-elbows stuff," he said). Saul Kent was there taking photographs of the whole procedure, and had been in and out of the operating room, putting new film in the camera, adjusting lights, and so on. Then, about halfway into the suspension, Kent walked into the room with an announcement.

"I just heard over the radio that Robert Heinlein died," he said.

It was a rather spooky moment. Heinlein was one of the true visionaries. He'd foreseen space colonies, cryonics, even a variety of the tiny robots ("Waldos," as he called them in a story) that Eric Drexler would reinvent some forty years later. Almost every person in the room could be counted as a great fan of Robert Heinlein's—Keith had actually named one of his daughters Virginia Heinlein Henson, in honor of the author's wife—but the strange fact was that it wasn't Robert *Heinlein* there on the operating table; it was some *other* Robert.

The people at Alcor couldn't fathom it, especially Keith Henson, who had tried to convince Heinlein, when the writer was still on the L5 Society's board of directors, to sign up for a cryonic suspension. In fact, so had Eric Drexler. Henson and Drexler had met with Heinlein at an L5 conference in San Francisco, and tried to tell him what a terrible loss it would be to science fiction, to science, to the world at large, if he should simply *let himself die,* once and for all and forever. But Heinlein hadn't agreed.

"If I could have been successful in talking him into the cryonics arrangements," Henson said later, "it would have been, *by far,* the most important thing that I had ever been involved with—ever at all."

But there was Henson at that very moment assisting with the suspension of this other Robert, the one he never knew. "It's got to be one of the strangest feelings, really an eerie feeling," he said. "This obscure television repairman from Florida, *he* was the guy going into the future and Heinlein wasn't. Heinlein was in a much better position than probably anybody else to understand what was going on, but he didn't follow through."

Afterward, Mike Darwin also gave some thought to the irony of the two Roberts dying within hours of each other.

"What an extraordinary and amazing situation," he wrote in *Cryonics* magazine. "An average, anonymous, middle-class man undertakes a desperate voyage across time and space to await rescue by physicians perhaps yet unborn, while the 'Dean of Science Fiction and America's foremost visionary' is cremated and his ashes scattered from a Coast Guard vessel. Reality *is* stranger by far than science fiction."

Not two weeks prior to the deaths of the two Roberts, another one of those great science fiction visionaries had also passed away, Clifford Simak. He had written about cryonics too, in *Why Call Them Back from Heaven?* and Mike Darwin had even spoken to him once, in 1977, at a science fiction convention in Silver Spring, Maryland.

"It was an interesting meeting," Darwin recalled much later. "Simak was at once both fascinated and, it seemed to me, a little repulsed by cryonics." Simak was a practicing Catholic and apparently thought that having just another mundane life on earth would be far inferior to what he could experience up in heaven. At any rate, he'd written in his book of finding, in heaven, "a better second life than Forever Center plans."

It was another unbelievable situation for the cryonicists, who could never comprehend why it was that most people did nothing at all even to keep themselves healthy, let alone to come back and live again. "This is the most ambitious project in the history of man," Art Quaife once said. "It's hard to understand why we don't have five billion customers."

Heinlein's death was the hardest of all to take, but Mike Darwin, for one, never thought they froze the wrong Bob.

"It's certainly true that our Bob was no Robert Heinlein," Darwin said. "But Bob had and still has something Heinlein hasn't a chance in the world of now: the prospect of immortality in an open-ended world of incredible possibilities. For he had the courage and the brains not to merely hear about The Door into Summer, but to actually step through it."

The temporary end result of a cryonic suspension is, one has to admit, rather striking, even sublime, in its own way. A few days after he deanimated, Bob was floating head-down in a capsule filled with cold, clear liquid nitrogen. If you opened the lid to the stainless steel capsule that held him you could see him down there, wrapped up in his bright blue shroud.

You'd open the Dewar tank's lid and a cloud of white vapor would pour out the top. You'd have to wait a minute or two for this to clear away, but then you could see the surface of the liquid. You could even touch it if you wanted to, very briefly. All you'd feel is a spike of cold.

To see inside you'd have to hold your breath—so that you didn't inhale the frigid vapors and freeze your lungs—and then you could bend down and lower your head into the remaining white mist. When you did that, why then, yes . . . you could actually see all the way down through ten feet of stuff that seems clearer than air and is certainly a lot colder; you could see all the way down to the bottom of the tank.

One day, so long as cryonics hadn't been made illegal in the interim, Mike Darwin will be down there himself, or at least his head will be. Darwin and most of the other Alcorians are going neuro; no one in his right mind, they think, is going to want his old body back. "Whole-body's just ridiculous," they'll tell you.

Saul Kent will be down there too, one day, along with his mother, supposing that the county coroner didn't find her first, "melt" her head, and "chop it up."

Keith Henson will be there, together with his wife Arel, their daughter Amber, and their friend Tim Leary. Bob Ettinger, the man who started it all, will be in his own capsule, his "cryostat," in Oak Park, Michigan, at the facility he built there long ago in the twentieth century.

But the others will be taking their Cold Sleep, as Heinlein told them to, in Riverside.

4

Omnipotence, Plenitude & Co.

When the first people were getting themselves frozen, back in the sixties and seventies, no one had any more than the dimmest notion as to what it would take to get you back up and running again. Actually there was a way in which this didn't really matter, because it wasn't going to be *your* problem: you'd be down there in the cryonics tank and in no condition to worry about resurrection day or anything else. Leave that little detail to others, to the Eternal Engineers. The important thing was that science and technology would be making their usual hubristic strides in the interim, so while you were sleeping your way through the decades, scientists, researchers, and advanced thinkers of every stripe would be learning about nature the way they always had, until finally the reanimation of frozen bodies was as routine as a heart transplant.

It called for a special brand of optimism to think this way, perhaps, but it wasn't really a matter of having blind faith in science. Or if it *was* a faith, it was the kind that Keith Henson had spoken of as "the faith of Goddard."

"Goddard *knew* from calculation that the moon was in reach," Henson once said. "There were only two things about Apollo that might have surprised him. It occurred much sooner than he thought it would, and he would have been dismayed that we didn't stay there."

The fact was, science was progressing at ever-increasing rates, and even at this early stage in the cryonics game, a few vague hints had already been offered as to the types of advances that would be necessary and sufficient to raise the dead. First off there had been Bob Ettinger's suggestion from back in *The Prospect of Immortality* about those robot surgeons of the future: "Huge surgeon-machines, working twenty-four hours a day for decades or even centuries," he'd said, "will tenderly restore the frozen brains, cell by cell, or even molecule by molecule in critical areas."

That was sketchy at best, but at least it was better than nothing. Later, in 1977, Mike Darwin came up with another idea, a more biological approach to resurrection. He proposed the notion of altering white blood cells so that they'd make repairs to damaged tissues and organs of all types. "If we start with something like a normal white blood cell and assume it could be modified in most any way," he said, "we could build an ultraminiature, self-reduplicating repair unit." Send a gang of these repair units into the veins and capillaries of a defrosting patient, and they'd seek out damaged cells, diagnose their troubles, and restore them to health.

As to where the money was going to come from that would one day pay for all these robot surgeons, self-reduplicating repair machines, and so on, that too was pretty much left up to the progress of technology, but again there was nothing implausible about this. Anyone could point to all sorts of examples of how even the most complicated technological devices started off costing a large fortune but then fell in price soon afterward to the point where you could buy them for next to nothing. Computers were of course the most obvious example. The first one, the ENIAC, had cost $400,000 in 1945 dollars, and thirty years later you could buy a hand calculator that did all the ENIAC could do, and more, for about $10. It was this kind of phenomenon that led Mike Darwin to speculate that the technology required to revive a frozen patient would cost "the equivalent of a $3 LCD wristwatch—or less."

Whatever it would cost, future revival was an obvious necessity. Ettinger himself (whose talents included making up atrocious puns, wisecracks, and Truly Immortal Poetry) once expressed the point in a piece of verse titled "The Man in the Can":

You've got to revive me
I need a live me
So some day defrost me
When it's not too costly
It's condition red
When you wake up dead.

Anyway, after a while it seemed that the real question was not whether a revival would be possible but why anyone in the far future would bother to perform one. Why would they try to resurrect someone they never knew, to whom they had no sense of personal obligation—a person, moreover, who would in all probability be grossly unfit, at least at first, for life in the century in which he or she would awake? For a long time many cryonicists didn't pay much attention to this question either, but then in the spring of 1989, Linda Chamberlain, one of the old-line cryonicists and a cofounder, with her husband, Fred, of the Alcor Life Extension Foundation, realized that the whole problem of reanimation had to be taken seriously. It made no sense to spend your life savings on getting frozen only to leave your resurrection up to the goodwill and altruistic sentiments of others. Her own idea was to start a self-help group called Lifepact, in which each member would promise to help unfreeze the next. Once the first person got thawed out, he or she would see to it that the next one got reanimated too, and so on down the line. There would be no problem with the *first* candidate, who would be revived by future scientists out of a sense of charity or curiosity if for no other reason.

Of course there were risks even here. After all, who or what would guarantee that any of the Lifepact members would keep their promise? What would guarantee they'd even *remember* it? There were no such guarantees; nevertheless, cryonicists rushed to join Lifepact, for the fact was that the whole idea of freezing and reviving the dear departed was becoming ever more reasonable. In the previous few years just precisely the right scientific insight, an entirely new way of getting control over nature, had appeared on the conceptual horizon: *nanotechnology.*

Nanotechnology would be a panacea of major and unprecedented proportions: it was a technology that would make just about *anything* possible, and this was meant quite literally. If it worked, nanotechnology would bestow upon human beings powers that had in earlier ages been thought to belong only to the gods. Specifically, it would give you, as Eric Drexler soon realized, *complete control over the structure of matter.* It would be a technology *so* powerful, *so* momentous in its effects, that raising the dead would be only one of its more minor achievements. Indeed, some scientists had already worked out nanotechnological schemes by which that very miracle might be accomplished. One of them was Ralph Merkle.

Merkle was one of the few mainstream scientific geniuses of the cryonics movement. A Stanford Ph.D. in electrical engineering, Merkle had appeared in *Time* magazine while he was still a grad student, in a story on encryption algorithms for unbreakable ciphers, one of which he'd helped invent. Later he worked at Xerox PARC, the Palo Alto Research Center, by any standard a world-class place for research and development. Merkle had once read Drexler's book, *Engines of Creation,* a nontechnical account of nanotechnology that told how little molecule-size robots, or *assemblers,* would perform just about any task required by humankind, virtually for free. The book had a chapter on cryonics and featured a scenario about how the assemblers could repair and revive a frozen body, eventually restoring it to an adolescent springtime glow.

It was an engaging account, and the whole program was entirely doable in Ralph Merkle's eyes, but still there was all the difference in the world between a popularization and a detailed scientific description of how it would work, especially down at the molecular level, where these nanomachines were supposed to be doing their jobs. Besides, Merkle was interested in the brain itself, the seat of consciousness and personal identity, and he wanted to know precisely how Drexler's assemblers would repair the brain cells that might have suffered freezing (or other) damage. No one else seemed to be working on the problem, so Merkle embarked on his own personal research program, the quest for a realistic brain repair methodology.

By the time he'd finished working it all out (this was in mid-1989), Merkle had decided that Drexler had been entirely right in his predictions: the assemblers *could* indeed do everything that Drexler had claimed they could do, making most of their repairs to the brain while the patient was still frozen. With the patient rigid and immobile at −196°C, a vast number of Drexler's little nanotechnological marvels would saw their way down through the frozen gray matter, survey the injuries done to molecule and cell, then put everything back to rights again, molecule by molecule, atom by atom. If the initial suspension had been done correctly (and maybe even if it hadn't), then the subject's memory, his personal identity, and his subjective feeling of being himself all ought to be fully present from resurrection day onward.

When the repair units had finished up, the patient would be thawed out, new blood would be pumped into his veins, and finally the subject would arise and walk, exactly as if he were a latter-day Jesus. It would be, quite literally, a resurrection of the flesh— except that all the miracles would have been performed by science.

*I*n December of 1959, when Eric Drexler was just four years old and a rather wee tot himself, Richard Feynman, the future Nobel Prize–winning physicist, best-selling author, and sometime bongo drummer, gave a talk at the annual meeting of the American Physical Society at the California Institute of Technology, in Pasadena. Feynman was one of those rare physicists who understood how actual flesh-and-blood objects worked out there in the real world; he didn't limit himself just to the quarks and sparks of the subatomic realm, as did many of his physicist brethren and sistern. His magnum opus, *The Feynman Lectures on Physics,* was a gigantic three-volume textbook that covered all the usual subjects, but there amid the discussion of tensors, vector potentials, and quantized paramagnetic states was a separate chapter on the ratchet and pawl. A whole chapter. The ratchet and pawl is the geartooth-and-catch mechanism that allows a shaft to turn only in one direction, as for example in a ratchet wrench. True enough, Feynman discussed the contraption only for illustrative purposes, by way of making broader points regarding entropy, disorder, and the irreversibility of physical law; nevertheless, one had a hard time imagining the average head-in-the-cloud-chamber particle physicist making any sense whatsoever out of a socket-wrench mechanism.

Anyway, Feynman's talk at the physical society meeting was called "There's Plenty of Room at the Bottom," and was a dazzling analysis of what human beings could actually do in the realm of the very tiny. "*Why cannot we write the entire 24 volumes of the* Encyclopaedia Britannica *on the head of a pin?*" he asked, and then he went on to explain how you could, at least in principle, do exactly that.

If you magnified the pin head twenty-five thousand times, he said, this would give you a working area equal to the total page area of the *Britannica*. In other words, if you took every page of the *Britannica* and spread them all out, then the total area covered by both sides of all those pages would be equal in square footage to the area of a pin head that had been magnified by a factor of twenty-five thousand.

But that was not putting the *Britannica* on a pin head. To do that you'd simply reverse the process: instead of enlarging the pin head to encompass the full-size encyclopedia you'd reduce the encyclopedia to fit on the pin. What you'd do is, you'd reduce all the letters of the alphabet and everything else in the *Britannica*— even including the little dots in the halftone illustrations—by the same factor, twenty-five thousand. Such a size reduction was physically possible, Feynman said, because even at that fine scale the smallest halftone dot would still be some thirty-two atoms across, containing within its area about a thousand atoms.

"So, each dot can easily be adjusted in size as required by the photoengraving, and there is no question that there is enough room on the head of a pin to put all of the *Encyclopaedia Britannica*."

That, though, was only the beginning. Indeed, it was next to nothing at all.

"Now, the name of this talk is 'There's *Plenty* of Room at the Bottom,'" Feynman said, "not just 'There Is Room at the Bottom.'" And then he went on to show how it was physically possible—meaning that it was allowed by the laws of nature, no miracles or magic were necessary to accomplish it—to put *more* than the *Encyclopaedia Britannica* on the head of a pin.

Much more! In fact, Feynman had worked out a method for fitting virtually *all of human knowledge* into a volume that was actually *smaller* than the head of a pin! He calculated that if you took all the books in the Library of Congress (at that time, nine million volumes), plus those in the British Museum Library (five million), plus those in the Bibliothèque nationale in France (five million more), and subtracted a few to eliminate duplications and threw in a few million more for good measure, you'd arrive at a figure of approximately twenty-four million volumes as the total number of books of interest in the entire world. He then argued that you could pack the contents of *all of those books* into a mass of metal that was *tinier* than a pin head. *Far* tinier!

Now any calm and reasonable person in the audience might well have been skeptical at such lunatic claims, but Feynman continued on in his matter-of-fact way to show precisely how you could do it. What you'd do is, you'd represent individual letters of the alphabet by means of dots and dashes, each of which would be about five atoms long. You'd transcribe the twenty four million books into this code and then write out the dots and dashes on the pin head, not only on the surface, but also on successive layers of metal underneath. Atoms are so small, and there are so many layers of them in the head of an ordinary pin, that in fact you'd have plenty of room.

"And it turns out," Feynman said, "that all of the information that man has carefully accumulated in all the books of the world can be written in this form in a cube of material $1/200$ of an inch wide—which is the barest piece of dust that can be made out by the human eye. So there is *plenty* of room at the bottom! Don't tell me about microfilm!"

Tour de force that it was, Feynman had something more in mind than merely entertaining his physicist friends in the audience. He thought there'd be some practical applications for this process, not only in the form of tiny machines and computers, but in the increased power over nature that atom-by-atom manipulation of matter would give you. Once you had a mechanism for moving atoms around one by one, he thought, it wouldn't be long before you'd have the ability to synthesize virtually *anything you wanted*. You'd be able to manufacture it *directly*, at the atomic level, just the way Mother Nature herself did.

"Give the orders and the physicist synthesizes it. How? Put the atoms down where the chemist says, and so you make the substance."

By the time the physicist figured out how to handle individual atoms, Feynman said, "he will have figured out how to synthesize absolutely anything."

And there it was, the Bashful Confession of Omnipotence. Sooner or later the scientist will be able to *synthesize absolutely anything.*

At the time Eric Drexler first got the idea for nanotechnology he hadn't yet heard of the "Plenty of Room at the Bottom" talk, and in fact he wouldn't read a transcript of Feynman's lecture until several years afterward. But Drexler had gotten some similar ideas on his own, in a self-imposed course of readings at the science library of the Massachusetts Institute of Technology, where he was a student.

When he first arrived at MIT Drexler's interests lay in space colonization. Since way back in high school he'd always had, if one may say so, rather big ideas. He wanted not so much to *understand* the world after the fashion of a theoretical scientist as to figure out ways of *doing things* with it, exploiting nature's laws for the greater glory of humanity. He never figured out why it was that other scientists were apparently so timid about doing the same thing. For example, there was the so-called Club of Rome report, *Limits to Growth,* in which the authors predicted that the world would run out of resources in a matter of mere decades. Drexler had read this early on, as a student in Monmouth, Oregon, where he grew up, and even then he could see plenty of things wrong with their whole scenario. The authors just didn't seem to appreciate what could actually be *done* with the world.

For one thing, they hadn't allowed for the development of any new technologies; rather, they'd imagined that mankind would be limited now and forevermore to the science and technology that was already on hand. But that was an absolutely ludicrous assumption to make in view of the way science and technology had always progressed. For another thing, the Club of Rome people postulated that the only resources available were those of the earth, but this was bizarre in light of the obvious fact that there was a whole universe out there chock full of nothing but raw materials. The Apollo astronauts, indeed, had already come back with several boxes full of moon rocks; they'd brought them back, of course, for their scientific value, not for commercial use, but *still,* how could anyone in their right mind say that we're limited only to *earthly* resources when we'd already gotten our hands on some *nonearthly* ones? It was even worse that the *Limits to Growth* authors assumed that we were limited to the resources on the *surface* of earth. "Our mines barely scratch the surface of the globe," Drexler thought to himself at the time.

How very different a picture you got if you saw mankind's proper realm of action as being the whole solar system. And indeed, why not assume exactly that? Science fiction writers had been doing so for decades, and half their scenarios had already come true, or had even been *exceeded,* what with the moon flights, the automated probes to Venus and Mars, and so on. Spaceflight, true enough, was still in its infancy, but the fact was, *there was a whole universe out there.*

Unfortunately, when he began to pursue these thoughts further, bicycling to the library at Oregon State University in Corvallis, twenty miles away from where he lived, Drexler found that there were a few practical problems standing in the way of immediate solar system colonization. The moon, for example, although it was the closest body to earth, was not all that rich in anything that human beings needed for life. It lacked water, and it didn't have much in the way of carbon, nitrogen, or hydrogen, whereas people, plants, and living things in general required just those ingredients in large volumes. And as for the inner planets, they weren't much better. "Venus is a hellhole," Drexler thought. "Mars is pretty distant and a worse place than Antarctica. So what's the point of going to Mars and Venus?"

Saturn's rings were a different story: they were made of ice, so Drexler began to think about mining them for their water content. But on the other hand, Saturn was almost halfway to the edge of the solar system.

Then, however, there were the asteroids. They were much closer to earth, most of them lying between Mars and Jupiter, and represented floating mother lodes of precious metals: iron, nickel, cobalt, platinum, even gold. Just one medium-size asteroid, Drexler estimated, would be worth some trillions of dollars in raw materials. There'd be problems getting out there and bringing it all back, but this was the very stuff of aerospace engineering, and at length Drexler decided that this would be the subject he'd study in college. It would be his way of helping to cope with the great problems of our time, a way of easing what the Club of Rome people had referred to as "the predicament of mankind."

So one of the first things Eric Drexler did when he arrived at MIT was look up people who were doing similar advanced thinking about the future of the species. An adviser told him to see Philip Morrison, the physicist, but Morrison, as it turned out, was not a big fan of space travel. Nevertheless he told Drexler about someone who *was* in fact working on the problem of space settlements, Gerard K. O'Neill, of Princeton.

That was in the fall of 1973, when Drexler was a freshman, and about a year before O'Neill would publish his *Physics Today* piece. Drexler got in touch with O'Neill, and events after that moved fast, so quickly that by the following May, Drexler, now all of nineteen years old, was in Princeton giving his first scientific talk, "Space Colony Supply from Asteroidal Materials," at the First Princeton Conference on Space Manufacturing. Following that there were the other Princeton conferences, the Ames Summer Study groups, and so forth, all of which Drexler attended, many of which he helped organize. Later on he worked with O'Neill on the Model I Mass-Driver, a contraption that was held together with epoxy. "I ended up with T-shirts full of the stuff," Drexler recalled later. "It coagulated in brittle patches which snapped and cracked the fabric. It was quite disgusting."

Drexler met Carolyn and Keith Henson at Princeton and joined the L5 Society, of which he was one of the first members. Naturally he had to go down there to Tucson, to "help out," and eventually he went the whole route, crawling around in the tunnels, feeding the goats, even editing an issue of the *L5 News*. "Carolyn was very busy, or had just had a baby, or some damn thing."

Indeed at this precise juncture it seemed about as certain as anything could be that Eric Drexler would wind up as one of the grand old men of the space colony movement—but then came his cosmic transfiguration. It was not a sudden thing, not exactly a eureka experience, it was just the germ of an idea at first, but gradually it developed into something bigger and bigger until finally he realized that this idea would in fact change absolutely *everything*. After nanotechnology, going into space would be *easy*, mining the asteroids would be *nothing*, even interstellar travel would not be out of the bounds of possibility.

"This is going to be the largest technological revolution that we've seen," he was to say later, "comparable to the invention of agriculture or industry."

*D*rexler got the idea for nanotechnology in 1976, roughly twenty years after Watson and Crick discovered the spiral structure of DNA. This was also the time that the genetic engineering business was just getting started, and the fact that the two disciplines were born simultaneously was by no means a coincidence. Genetic engineering, after all, involved making changes at the bottom of things, modifying DNA to make it serve your own purposes: it was a way of forcing the chromosomes to do what *you* wanted them to do rather than what *nature* had programmed them to do. This meant that scientists were learning how to operate nature at its most fundamental levels: they were reprogramming nature's own tiny robots, DNA molecules. After a while Drexler got to wondering whether human beings could actually learn to *make* such programmed molecules.

"Sometime in 1976," Drexler recalled, "I started thinking seriously about what you could build if you could design protein molecules and other biomolecules. I could see from the literature that there were all these mechanical and electronic widgets inside cells, that these things were synthesized chemically by the cells, that they spontaneously assembled inside them, or even in the test tube: you mixed the parts together and through selective stickiness they grabbed on to each other to make complicated little devices. And I asked myself, Well, what if we could do things like that?"

What, for example, if you could make complicated little devices like *robots*, tiny mechanical marvels that were roughly DNA-size? These man-made robots—Drexler called them *assemblers*—would be so small that they'd be able to manipulate individual molecules of matter, or even *atoms themselves,* one by one. Supposing that you could control the assemblers by means of internal programs, then they could accomplish absolutely amazing feats. They'd be able to place atoms together in any chemically allowable configuration, synthesizing for you whatever substance you wanted. They'd be able to position molecules of matter in any structurally stable conformation, allowing you to build virtually anything at all. What, indeed, *couldn't* they do?

"With the assembler," Drexler said, "you'd be able to take reactive molecules, put them in specific places, and control chemical synthesis to build up complex structures. All the operations you're performing are familiar to organic chemists, you're just getting a lot more control over where they occur, and maybe on occasion doing somewhat novel things by being able to push—with an electric current, perhaps—to make something happen that wouldn't happen otherwise."

Having an assembler would be like having some motorized DNA at your fingertips, except that instead of producing only living organisms, these assemblers could fabricate any possible physical structure. They'd be no more and no less than *universal building machines.*

"That very rapidly began to look like something very big and important. Because if you have this very general ability to stack up atoms in complex patterns, well, then you can make essentially anything that's physically possible."

And there it was again, that Bashful Confession of Omnipotence. You could *make anything that's physically possible.*

*I*n his "Plenty of Room at the Bottom" talk, Feynman described a simple mechanical procedure for the construction of increasingly tiny machines. The trick was to make a device that would make an identical copy of itself, only smaller. This small-size unit then makes an even smaller copy of itself, and you'd keep on going like this until you ended up with the smallest physically possible machine.

The original machine, Feynman said, would be controlled by a human operator through a system of master and slave hands. You'd put your hands inside a pair of gloves that followed what you did and duplicated your movements in a second set of hands, the slave hands. If the slave hands were only a quarter of the size of the originals, they'd do everything on a proportionately reduced scale; so if the master hands went through the motions of building a full-size lathe, for example, then the slave hands would build an identical lathe a quarter as large. For this to work you'd first have to equip the slave hands with a set of quarter-size parts and tools to work with—tiny nuts, bolts, screwdrivers, wrenches, and so on—but once you did that the slave hands could build almost anything at a quarter scale, including the tools to make even smaller parts for even smaller machines.

You could go on like this, Feynman said, until you'd produced "one little baby lathe four thousand times smaller than usual." But why stop at just *one* such lathe? Why not wire up the master hands so that they'd control *ten* quarter-size slave hands, so that you'd be producing *ten baby lathes* in a single step? You could continue the process exponentially until at the end of it your master hands would be controlling *a billion tiny baby lathes.*

Instead of building lathes, however, the slave hands could be made to build anything you wanted them to build: whole assembly lines, entire factories, and you could do it for almost *nothing at all.* Said Feynman, "It doesn't cost anything for materials, you see. So I want to build a billion tiny factories, models of each other, which are manufacturing simultaneously, drilling holes, stamping parts, and so on."

In fact you could get even *smaller* than that. There'd be no inherent stopping point until you got right down to the level of atoms. In theory, you'd be able to manipulate *individual atoms,* one by one.

"The principles of physics, so far as I can see," Feynman said, "do not speak against the possibility of maneuvering things atom by atom. It is not an attempt to violate any laws; it is something, in principle, that can be done."

And then Feynman reached the end of his lecture.

"It is my intention to offer a prize of one thousand dollars to the first guy who can take the information on the page of a book and put it on an area $1/25,000$ smaller in linear scale in such manner that it can be read by an electron microscope."

(Whoever did that would of course be able—if he had the time, money, and patience—to write out the *Encyclopaedia Britannica* on the head of a pin.)

"And I want to offer another prize—if I can figure out how to phrase it so that I don't get into a mess of arguments about definitions—of another one thousand dollars to the first guy who makes an operating electric motor—a rotating electric motor which can be controlled from the outside and, not counting the lead-in wires, is only ¹⁄₆₄ inch cube.

"I do not expect that such prizes will have to wait very long for claimants."

*A*s Eric Drexler conceived of it, you wouldn't have to go through any such progressive downsizing as Feynman had described. Drexler thought you might be able to program nature's own biological devices—proteins, for example—so that they'd help put together the first wave of assemblers, after which the assemblers could be programmed to make identical copies of themselves. Such self-copying assemblers he called *replicators*.

The replicator was even *more* like motorized DNA than the initial assembler was. DNA's main goal in life was to copy itself, and the same thing would be true of Drexler's mechanical replicators. They wouldn't *look* like DNA molecules, which were long, loopy, spiraling strands, nor would they *work* the way DNA molecules did, by splitting and splicing; nevertheless, both of them were essentially small self-reproducing machines.

Drexler's replicator would be smaller than a cell, smaller even than a cell's nucleus. It would be *so* small that its surface would be knobbed and bumpy, and these bumps would be individual atoms. These atoms would form the replicator's basic components, drive shafts, gears, bearings, motors, housings, and so on. Once it was put together and operating, the assembler would work much like an industrial robot, reaching out for other components (atoms or molecules), positioning them according to a final plan or blueprint, doing this again and again, like a bricklayer, only at speeds on the order of a million bricks per second.

One time Drexler calculated how long it would take a replicator operating at that speed to make another complete copy of itself. "Working at one million atoms per second, the system will copy itself in one thousand seconds, or a bit over fifteen minutes—about the time a bacterium takes to replicate under good conditions."

That was not especially fast. "If this were all replicators could do," he said, "we could perhaps ignore them in safety. Each copy, though, will build yet more copies."

This gave you progressive doubling, an exponentiating process that could get out of control very quickly. "At the end of ten hours, there are not thirty-six new replicators, but over sixty-eight billion," he said. "In less than a day, they would weigh a ton; in less than two days, they would outweigh the earth; in another four hours, they would exceed the mass of the sun and all the planets combined—if the bottle of chemicals hadn't run dry long before."

This was where nanotechnology's *leverage* came from: not only did the initial replicator make a copy of itself, *so did each of its descendants*. If you could control a race of such tiny mechanized men, you could, at least potentially, *control the world*.

Not that Drexler had any such ambitions. His aspirations weren't political but scientific: he wanted control over matter at its finest levels. "Pretty soon you'd have a very powerful technology," he said, "one that essentially gave you complete control over the structure of matter."

And what was this but yet another Bashful Confession: *complete control over the structure of matter*. This would be such a consummate sway over nature that you'd be able to perform what might at first seem to be impossible tricks of molecular manipulation. You'd be able to synthesize fresh beef, for example, without benefit of cows. This miracle would be performed by the *Meat Machine*, also known as the Cabinet Beast.

The Meat Machine would be a box containing a waiting array of programmed assemblers. You'd open the box, shovel in a quantity of cheap raw materials—some dirt, straw, grass clippings, or whatever—then close the box and let the assemblers ply their trade, which is to say that they'd break down certain chemical bonds and make others, all according to a plan. After a while, you'd open the box and out would roll a wad of fresh beef.

Unbelievable, of course, but the fact was that the Meat Machine would only be doing what cattle themselves did as they turned water, grass, and sunlight into meat. And on second thought a person had to wonder which of the two ways of making beef was inherently more incredible, the cow's or the Meat Machine's.

What was uncanny about the *cow's* way of doing it was that the animal did everything *automatically*. It just sat there chewing its cud, and all by itself grew piles and pounds of beef on the hoof. How believable was *that*, really?

For that matter, so far as credibility went, a critical observer could easily get to wondering whether DNA's way of producing *anything* was really all that believable. Who'd ever have thought of it, that you could produce something as complex and differentiated as a living person from progressive divisions of *a single starting cell?* Who could expect that all those millions of lines of genetic code could be copied time and again, millions of times, with perfect accuracy? Who in their right mind would have believed, before the fact, that DNA replication could ever possibly work? Nobody. Nevertheless, it did.

And it would be the same way with Drexler's assemblers. It was hard to accept, at least at first, that they could ever possibly perform as advertised. But why wouldn't they? They'd *have to,* because there was already a proof-of-concept in nature itself.

"Proof for self-replicating systems of molecular machinery exists in the form of bacteria," Drexler once said. "Anytime someone makes yogurt, he's demonstrating that self-replicating molecular machines work. I can't see how to construct an argument against these ideas that does not also deny things we know exist."

Indeed, the more you thought about this, the clearer it became that nanotechnology would be the most stupendous panacea in the history of the species. You came to realize that once nanotechnology got going there'd be no more *hunger:* assemblers could make more food than you'd ever want to eat.

There'd be no more *poverty:* assemblers would manufacture all the material possessions you could imagine—everything from cars to spaceships—out of the cheapest ingredients.

There'd be no more *human labor:* assemblers worked essentially for free, just like living cells did. "When trees grow," Drexler said, "the manufacture of that wood does not require human labor. The energy comes from the sun, materials from the atmosphere. In a similar fashion, it should be possible to make seeds that, when given the right nutrients, will 'grow' to make almost anything physically possible."

There'd be no more *large corporations:* the assemblers would do all the work.

There'd be no more *disease:* assemblers equipped with medical expert systems would diagnose illnesses and make the necessary repairs.

In fact, given the miraculous nature of what the assemblers *could* do, Drexler found he had to spend some time telling the increasing numbers of his readers and listeners that the assemblers could *not* in fact do literally *all* things.

"Nanotechnology will not make everything possible," he said. "No matter how you arrange atoms, some things cannot be done. Natural law—whatever it may be—determines what matter is and what it can do. It will set bounds to the strength of materials, the speed of computers, and the rate of travel."

Even nanotechnology couldn't give you faster-than-light travel. Nevertheless, within the domain of what *was* possible, it seemed that nanotechnology could accomplish virtually all things, quickly, cheaply, and without your having to lift a finger.

A few months after Feynman offered his thousand-dollar cash prizes, William McLellan, a Caltech physicist, collected one of them. After two-and-a-half months of lunch-hour work with a microscope, toothpick, and a watchmaker's lathe, he'd actually managed to build a working electric motor that was less than $1/64$ inch on a side.

Feynman handed over the money, but then he got worried. This was before he'd won his *own* prize, the Nobel, and he was already strapped for cash. After all, he'd gotten married and bought a house, all of this on a college professor's income. He got so worried, in fact, that he tried to get the people who were working to collect the other one thousand dollars to *cool it*.

"This, then, is a public appeal to all inventors who are now at work trying to write small and collect the Second Feynman Prize— TAKE YOUR TIME! WORK SLOWLY! RELAX!"

As it turned out, he needn't have been so concerned. Nobody succeeded at the second problem until November of 1985, when

Tom Newman, a graduate student in electrical engineering at Stanford, wrote out the first page of *A Tale of Two Cities* at the required 1/25,000 scale reduction. He formed the individual letters using a beam of electrons about one five-millionth of an inch in diameter, and then got an image of the page as a whole by the use of a scanning electron microscope, just exactly as Feynman had envisioned. The image was reproduced, quite readably, in the January 1986 issue of the Caltech journal *Engineering and Science*.

Separately, and at about the same time Tom Newman was writing out his nanoscale *Tale of Two Cities* paragraph, others had achieved even finer levels of atomic manipulation. In fact, it seemed that scientists had finally reached what Feynman had spoken of as "the bottom," smaller than which it was not possible to go. They got there with the scanning tunnelling microscope.

The scanning tunnelling microscope, or STM, had been invented in 1981 by researchers working at the IBM Research Labs in Zurich. When reduced to essentials it was nothing more than an ultrasharp needle—much like a phonograph stylus, only finer—that could be maneuvered to atomic tolerances. In addition to reading off the atomic hills and valleys of a given surface, the needle could also make atomic-scale changes to it. The STM seemed to give you in one fell swoop exactly the fine-grained dexterity that Feynman had talked about, only without the long, intermediate series of master and slave hands.

In fact, six years after it had been invented, scientists had reached the end of the road with the scanning tunnelling microscope. Engineers working at Bell Labs had used it to deposit *a single atom* on a flat surface.

As they wrote in the British science journal, *Nature:* "We believe this to be the smallest spatially controlled, purposeful transformation yet impressed on matter and we argue that the limit set by the discreteness of atomic structure has now essentially been reached. Man can now manipulate a few chosen atoms for his own purposes."

*O*nce word of nanotechnology got around, cryonicists were surer than ever that they had it made all the way to the end of time. It was only a short step, after all, from *complete control over the structure of matter* to *complete control over human biology,* and Drexler himself was not slow to make the connection. There was a whole chapter in his book *Engines of Creation* devoted to cryonics—"A Door to the Future"—making all of it seem quite nonflaky and reasonable. Indeed, one of the first applications Drexler had come up with for his little nanotechnological marvels was a frostbite cure.

Frostbite was, of course, a tailor-made condition for treatment by automated cell-repair mechanisms. The structure of the cell would be preserved by the very cold that damaged it in the first place. The assemblers would simply invade the damaged cell, make the necessary alterations, and get the whole thing operating again. In fact, once the structures had been repaired, they'd probably start functioning by themselves, automatically, just like the defrosted cat brains, the defrosted human embryos, and so on.

The extension to cryonics was then obvious. Drexler, who had initially thought that the freezing damage would be too great to make cryonic suspension survivable, had decided to look at the whole matter again, so he picked up a copy of Ettinger's book, *The Prospect of Immortality.*

"Ettinger had the idea that something like assemblers would be possible," Drexler said, looking back. "An explicit part of his argument was that at some point in the future there will be fabulous machines able to repair things molecule by molecule. He referred to 'huge surgeon-machines,' rather than extremely small ones, but still the basic idea of molecular repair was there, and it was fundamental to the cryonics idea."

Drexler was impressed enough by Ettinger's talk of molecular repair machines to write to him and ask if he'd been aware of Feynman's "Room at the Bottom" lecture or of the notion of small-scale manipulation of matter in general. As it turned out Ettinger hadn't, so here was an area where Drexler could do some pioneering work on his own. The result was that *Engines of Creation* contained what was probably the best scientific case for the idea of cryonics that had ever been written, which is why, a couple of years later, Alcor's attorneys asked Drexler to furnish them with a technical declaration in the Dora Kent case.

Drexler didn't like getting involved with cryonics any more than he had to: it was bad enough coming out with his nanotechnology stuff, let alone being associated with anything as crazy as "reviving the dead," but he felt that this time he had to make an exception. "That was a case," he said later, "where a woman's life was at stake."

Indeed, although her head had been cut off and was now resting in a tank of liquid nitrogen at parts unknown, Drexler did not regard Dora Kent as "dead" in any true sense of the word. She was rather in "biostasis," a term he had coined in *Engines of Creation,* where he'd defined it as "a condition in which an organism's cell and tissue structures are preserved, allowing later restoration by cell repair machines." Soon enough this usage, and other word coinages like them, had worked their way through the ranks of the cryonics movement. Much of this new rhetoric was the work of Brian Wowk, who had written what was to become one of the movement's more celebrated articles, "The Death of Death in Cryonics."

In the article, Wowk asked his fellow cryonicists some pointed questions: "How often have we struggled with impressions that cryonics is a *sacrilegious, ghoulish, or Frankenstein-like* practice when we try to explain the concept? How often have we had the impossible task of trying to overcome the notion that cryonics entails supernatural *resurrection* when we try to explain its scientific foundations?"

The root of the problem, he claimed, was the common perception, held even by some cryonicists themselves, that people in suspension were "dead." Wowk insisted, to the contrary, that *"cryonic suspension patients are not dead in any meaningful sense of the word at all."*

Wowk's reasoning here was that "death" referred to *irreversible* loss of life. This was inappropriate in cryonics, he argued, because if suspended patients could in fact be brought to life again in the future, it followed that they were never truly dead to begin with. Not that it would be correct to regard them as "alive" during this time, either. (Saul Kent , though, had once claimed that it would be correct to regard the suspended as being *"potentially* alive.")

Rather they were in some intermediate state, a separate biological category for which a new term ought to be provided and reserved. Within cryonics, the substitute term of choice had long been "deanimation," but here Wowk confessed that "'deanimation' has always struck me as vague, crude, contrived, and in fact like just another name for death. I would like to suggest some more precise alternatives."

He suggested several: *ischemic coma, ametabolic coma,* and *biostatic coma*. These, he said, were medically precise terms that accurately described the state you'd be in once you'd . . . "died," "deanimated," or, as Wowk himself put it, "required suspension." And then there was the term *in suspension* itself. "The term *in suspension* (with any luck) will gradually replace 'being dead' as a social designation for cryonics patients."

There was indeed something to be said for all this. Ettinger, after all, had been pointing out for a long time that people who have just "died," in the clinical, medical sense, were in fact "99 percent still alive." They were only "*slightly* dead," he claimed.

The truth of this observation was, if anything, *over*proved by the ease with which doctors in hospitals, emergency rooms, shock trauma units, ambulances, and elsewhere routinely resuscitated people who had no heartbeat or respiration and who were therefore genuinely "dead" in the clinical sense. Indeed sometimes patients who were both clinically *and* legally dead had spontaneously revived, as happened, for example, in April 1989 to an eighty-two-year-old Holyoke, Massachusetts, woman by the name of Helen Francoeur who one morning was found "dead" in her apartment. The medical examiner, Dr. William J. Dean, was called and, finding no signs of life ("There was no pulse," he said later. "She was cool and there was no heartbeat that we could hear"), pronounced her dead on the spot and ordered an autopsy.

But then, on her way to the morgue, she revived, creating, according to one of the attending paramedics, "a real embarrassment to the medical community. I really wish this had never happened." (The Alcorians got a big kick out of this one. "Alcor isn't the only organization saving people from coroners," one member said.)

Later, Steve Harris, the Alcor member who was an M.D. and a resident at the UCLA Medical Center, published an article in *Cryonics* called "Binary Statutes, Analog World," in which he offered several select arguments that there was in fact *no sharp dividing line* between life and death. "Human beings come into existence a little bit at a time, as the abortion issue has taught us," he said. *"Humans go out of existence in the same way."*

All of which showed that Brian Wowk had at least some measure of justification for wanting fresh nomenclature—such as *in suspension*, or *in ischemic coma*—to refer to frozen cryonic patients. Anyway, his "Death of Death" article precipitated a debate within cryonics that continued on for months afterward.

Thomas Donaldson, one of Alcor's resident medical experts, said, "So long as I am frozen and revived it doesn't concern me whether my suspension is referred to as 'frozen storage meat' or 'cryostasis.'"

Hugh Hixon, another longtime cryonicist, agreed with Wowk that "cryonic suspension patients are not dead, in any meaningful sense of the word." Hixon claimed, nevertheless, that Wowk's "ischemic coma" terminology was not really an improvement. "*Coma* already has a well-defined medical usage," Hixon said, "and is rooted in the concept of sleep. Cryonic suspension is not sleep, but something new under the sun."

Eric Drexler, who was by this time speaking of cryonically suspended patients as suffering from "severe, long-term, whole-body frostbite," wrote a letter to the editor of *Cryonics* in support of Wowk's proposal. "To call [frozen] people dead is an abuse of language and sense," said Drexler. "It is misleading, upsetting, and destructive."

But finally Jerry Leaf, Alcor's chief surgeon, weighed in with what amounted to an insuperable objection to Wowk's *ischemic coma* jargon. Agreeing with Hixon, Leaf stated that *ischemic coma* already had an accepted medical meaning and it didn't mean "dead," it meant that a patient had lapsed into a coma as a result of an obstruction of the blood supply to the brain. But the fact was that plenty of ischemic coma victims later recovered from their condition and went home.

"If a novice cryonicist starts telling a physician that Alcor wants to take charge of a patient when ischemic coma occurs," Leaf said, "the cryonicist may find himself being escorted off the grounds of the medical facility by one of the security guards."

There was no immediate agreement on what was to be the new buzzword of choice for the frozen departed, but Wowk had achieved at least *one* thing with his "Death of Death" arguments: the "D" word was no longer used by any self-respecting cryonicist to refer to those in suspension; it was reserved for those who had gone all the way to being *Really and Truly Dead*.

*F*ar more important than terminology, though, was the question of exactly how cryonic patients would be revived. Drexler had presented his own resuscitation scenario in *Engines of Creation,* and it was entirely plausible as far as it went. Resuscitation wouldn't be at all difficult, Drexler said, once you realized that his tiny programmed assemblers would be doing most of the work. The assemblers would have vast medical knowledge programmed into their nanocomputers so that they'd be able to spot what was wrong with any given cell, and then provide the necessary healing services.

The nanomachines would do all this, Drexler suggested, while the patient was still frozen. His scenario was that they'd enter the chest cavity and clear out the blood vessels and capillaries. Fluid would be pumped through the circulatory system, flushing out the veins and carrying in a new wave of cell repair machines and supplies. At that point the nano cell repair machines would start to inspect the frozen biomolecules, one by one, and do whatever was necessary to cure them. "When molecules must be moved aside," Drexler wrote, "the machines label them for proper replacement. Like other advanced cell repair machines, these devices work under the direction of on-site nanocomputers."

At length the patient would be warmed up. The cells revive and begin to work. A fresh blood supply is grown from the patient's own cells and is put into circulation. The heart starts beating. The patient returns to life, though not yet to consciousness, because he's now asleep, as if under deep anesthesia.

The nanomachines withdraw from the body, healing any remaining wounds on their way out. Machines still trapped inside individual cells spontaneously disassemble into harmless components, which later emerge as waste or are metabolized as nutrients.

"As the patient moves into ordinary sleep," Drexler wrote, "certain visitors enter the room, as long planned.

"At last, the sleeper awakes refreshed to the light of a new day—and to the sight of old friends."

A couple of years after Drexler published this account, Mike Darwin wrote out his own resuscitation narrative, adding in a few more details. He described how the repair machines would take cells whose structures were almost totally obliterated and restore even *them* to proper functioning. They'd do this by getting copies of the correct DNA sequences from the healthy cells that remained, then loading those sequences back into the damaged cells. After this new infusion of lifesaving genetic information, those cells would function just as they had before.

Other nanomachines would forcibly regress the frozen body's biological age back to the desired state of youth. If necessary, as in the case of a neurosuspension, the nanomachines would grow a fresh body from the patient's old cells. ("Imagine it like this," Keith Henson once suggested. "You drop the head into a bucket of nanomachinery and a new body grows out of the stump of the neck.")

The total repair process, in Mike Darwin's account, takes "a little over a year," after which the patient awakens in his hospital bed. As before, the patient's wife and family, having already been re-animated, are there standing by.

"A familiar voice calls out his name. Instantly there is recognition. It is his wife. But she is not as she was. She is young and beautiful again. More beautiful even than he remembered. An instant before he was trapped in a dying body. Now, he is alive and well and looking into the eyes of someone he loves."

That was getting to be the conventional resuscitation scenario, waking up among spouse, family, and friends. But others, who evidently had different reasons for coming back, viewed all this togetherness with some distaste. Bob Ettinger remembered once telling a cryonics prospect in glowing terms about how she and her husband would be resurrected together.

"In that case, count me out," she said. "I don't want him when I'm thawed, I want five hundred years of free love."

None of these scenarios, as appealing as they were, addressed the specific problem of brain repair. This was unfortunate, because that was exactly what the whole cryonics dream rested upon. Ralph Merkle was noticing this embarrassing lacuna as he wondered whether he personally should sign up to be frozen.

Merkle had been raised in Livermore, California, home of the Lawrence Radiation Laboratory. His father, a physicist, was associate director of the lab and in fact had been next in line to become the director of the place when he died of cancer at the age of forty-seven. Merkle was only fourteen at the time, and the event made an impression on him, for it was a prolonged and painful death. His father had planned for him to be a medical doctor and had started teaching him anatomy and biochemistry, and after his father's death Ralph decided he'd take on the question of life extension as a kind of private research project. After all, plenty of other species lived for a lot longer than humans did—land tortoises, for example, or redwood trees, or even *ivy,* the lowly climbing vine. Why should *they* have thousands of years or more of life, some of them, while humans were given a mere three score and ten, if that?

So after he got his Stanford Ph.D. in electrical engineering, which was then the academic term for computer science, Merkle went back for some postgraduate courses, those in the neurosciences especially, and began to learn what he could about the brain and its software. Maybe there'd be some way to extract the information from inside a person's brain so that the person could live on in some fashion after his mortal body's death.

While he was in the middle of this research Merkle attended a lecture by Eric Drexler. "The basic idea—that of programmable self-reproducing devices—certainly seemed plausible," Merkle said later. "After reading Feynman's article, which was mentioned by Drexler, I found the logic quite inescapable."

Merkle then got a copy of *Engines of Creation.* "The one thing I found surprising on reading Drexler's book was his claim that tissue could be repaired molecule by molecule. If true, this was of course quite wonderful."

It was wonderful because, among other reasons, it would allow you to repair the brain itself. But how? Cryonics, Merkle decided, essentially came down to two questions. One, could you revive a frozen brain? Two, would the individual's memory revive along with him? If either of these was *not* possible, then the person was just as good as dead.

Merkle approached the problem by taking a negative tack: Can I prove that memory revival *won't* work? So he tried to imagine circumstances in which human memory would not in fact be preserved after freezing and thawing, but all the scenarios he could think of turned out to be quite unrealistic. The fact was that human memory had a physical basis to it: when you memorized something—a telephone number, for example—your brain was physically changed in the process. Human memories lasted through sleep, sickness, anesthetics, drugs, alcohol, all manner of physical and mental abuse, so why should it be any different with freezing?

But on the other hand there was the obvious fact that the processes of freezing and thawing caused physical injuries to cells, which meant that freezing would *interfere* with memory, which in turn meant that if cryonic suspension was going to work there had to be a way of undoing those injuries.

Merkle calculated the total number of molecules in the human brain: 2×10^{23}. Assuming that it would take three years (which Merkle thought was a highly conservative figure) to inspect and repair every last one of these molecules, it would follow that you'd need a fleet of 1.8×10^{16} cell repair machines to effect a total molecular overhaul of the brain. This was a considerable number of assemblers, vastly exceeding the number of stars in an ordinary galaxy, but not a quantity difficult to produce once they'd started in with their progressive exponential doubling: you could have a ton of assemblers, Drexler had shown, in a matter of hours, whereas the amount needed by Merkle was far less. In fact Merkle had calculated that if each machine weighed 10^9 amu (atomic mass units), then the total combined weight of all 1.8×10^{16} cell repair machines would be some thirty grams, or about *one ounce*. That was how small those assemblers were.

Manufacturing the repair machines was no problem. The main challenge was to come up with a way of having those cell repair machines inspect and rebuild each individual molecule of which the brain was composed, while the brain itself was still frozen.

Unfortunately, analyzing the molecular structure of a frozen object would be impossible without physically separating molecules from one another, and the only way of doing this was by making spaces between the molecules—in other words, by introducing networks of cracks throughout the object. Accordingly, Merkle assumed this was the way the cell repair machines would work, by deliberately inducing cracks among the brain cells. Thus, what we'd have on our hands now would be not only a dead brain, but a frozen dead brain, and in fact a *pulverized* frozen dead brain, one whose gray matter had been turned into a powder a lot finer than was ever made for cosmetic use.

But one shouldn't be put off by this network of brain cracks, Merkle said, because "cracks made at low temperatures are extremely clean, and result in little or no loss of structural information. As an example, when a jigsaw puzzle is assembled the picture on it is easily recognizable, despite the 'cracks' between the pieces. The loss of information involved in dividing a picture on a piece of cardboard into pieces is quite small."

Indeed, all this systematic brain cracking would have been an alarming prospect were it not for the fact that as long as you tagged the location of each molecule, you ought to be able to put all the pieces together again, just like a giant jigsaw puzzle. Fortunately, tagging and remembering the location of each brain molecule would be no problem at all when you had 1.8×10^{16} cell repair machines on hand to do the work.

As for the feasibility of putting the pieces together again, Merkle noted, after Drexler's own fashion, that a proof-of-concept already existed. "When you consider that you were built by bringing each and every molecule to your brain in the circulatory system, this should not appear too infeasible."

But there was one more matter. Just as surgeons make slight improvements to the human body once they've cut you open (the appendix is often removed during cesarean birth), so too might Drexler's assemblers improve your brain as they're churning their way through it. Of course the repair machines would make only those changes that were *safe*. "For example, moving subcellular organelles within a cell would be safe," Merkle said, "because such motion occurs within living tissue. Likewise, gently pushing aside tissue to open a small space should be safe. Indeed, some operations that might at first appear dubious are almost certainly safe."

Ironically, one unintended consequence of Merkle's scenario had been anticipated by Ettinger long before, in his "Penultimate Trump" short story about the wealthy H. D. Haworth. Haworth had suffered the indignity of having his thoughts read out, put on file, and published. Given the physical basis of memory, such mind reading would of course be open to the cell repair machines that restored your brain to health, as Merkle himself was well aware.

"This does pose a significant risk," he admitted. "The implications for personal privacy are rather severe. Hopefully, the use of such mind-reading technology can be limited to cases of obvious need—for example, a person suspected of murder or the like. There will, however, be significant pressures to extend its use."

A small enough price to pay, perhaps, for the resurrection of the flesh.

Strangely enough, once nanotechnology got going, mankind would have regained Paradise, the Garden of Eden, by committing the very sin, hubris—trying to be like the gods—that originally got us thrown out of the place. We'd have acquired godlike powers and attributes: immortal life, complete control over the structure of matter, vast material riches, and so on, and all of it would be virtually without cost in raw materials or labor, all having been done by Eric Drexler's assemblers. After that, things will finally be just as they were in Paradise, when Adam and Eve had peace and plenty and freedom from labor, and the concept of death was completely unknown. All this plenitude will rain down upon us automatically, like manna from heaven.

It seemed too good to be true, even to Eric Drexler. It seemed as if there had to be a catch here somewhere, a punishment lying in wait for all this good fortune. And indeed there was: simply put, you had to pay a price for all that plenitude.

The point had been realized way back in the Middle Ages, when theologians tried to figure out why it was that an infinitely good God nevertheless allowed evil to exist in the created world, things like droughts and floods, epidemics and plagues, earthquakes and hurricanes, and above all, human beings, which were entities possessed of free will and an inordinate propensity for wrongdoing.

The answer was clear: God was so bountiful, so generous, so excessively overflowing in his creativity, that he wanted the world to *lack nothing*. He wanted all possibilities to be realized, but by definition *all* possibilities meant the bad along with the good; so from the supposition that God was infinitely prolific the theologians came to the conclusion that evil had to be a part of the world. If there were no evil in the world, then at least one thing would be lacking in God's creation—evil itself.

And so it was with Drexler's replicating assemblers. The good part about them was that they made virtually all things possible, and for free. They replicated of their own accord and did the world's work without recompense. (Drexler once described how his assemblers, working by themselves in a vat of fluid, would "grow" a new rocket engine. "It is a seamless thing, gemlike," he said of the newly grown rocket motor. "Its creation has required less than a day and almost no human attention.")

All that was for the good. But the bad side was, what would happen if those madly replicating assemblers ever . . . GOT OUT OF CONTROL?

The Meat Machine alone was enough to make you wonder. The advantage of course was that you'd get an unlimited supply of free beef and you wouldn't even have to kill animals to do it. No longer would you have to be a vegetarian, because now you could get meat from a *machine*.

All that was the good part, but what if the meat machine *just went on working*? After all, the nanotechnological assemblers worked of their own accord, just like bacteria did. *What if you couldn't stop them?* What if they took it into their tiny heads to keep on replicating *without limit*, churning out meat come what may?

Drexler had already calculated that from a single replicating assembler, you could get sixty-eight billion assemblers by the end of a ten-hour period. "In less than a day, they would weigh a ton," he'd said, "in less than two days, they would outweigh the earth; in another four hours, they would exceed the mass of the sun and all the planets combined."

That presented an entirely *different* picture of what it might be like to live in a home equipped with a Meat Machine. It was the picture of a modest suburban house—nay, of a gargantuan suburban mansion, having been built effortlessly and without cost by Drexler's little nanotechnological marvels—suddenly and without warning bursting forth at the seams from an overproduction of *meat,* from *the Meat Machine,* which is out there in the kitchen . . .

(*thumpa, thumpa, thumpa*)

. . . *running amok!*

Huge assembler-built slabs of beef come muscling out of the mansion and start lurching their way up the street, gobbling up all the other suburban mansions and assembler-produced sports cars in their wake. Soon it's nearing the city limits . . . this gigantic pulsating, rampaging mass is slurping its way into Chicago.

Chicago!—Stormy, husky, brawling; city of the big shoulders—and now all it is, is a *Big Shoulder of Beef!*

Meanwhile, from the other end of town comes an oversupply of assembler-produced rocket engines! These seemless, gemlike diamond-and-sapphire beauties have been pouring out of the factory for days and have by now turned the surrounding countryside and everything in it—shopping malls, theme parks, cryonics laboratories—into *spare rocket parts!*

This is what you get for your hubris! *This* is what you get for imagining that mankind—mere mortals—can attain anything as grand as *complete control over the structure of matter!*

Actually, Drexler had seen quite early on in the game that out-of-control assemblers might turn out to be a problem. In fact, the specter of replicators running amok soon went under the heading of *the gray-goo problem* in nanotechnology circles.

"The gray-goo threat makes one thing perfectly clear," Drexler had written in *Engines of Creation.* "We cannot afford certain kinds of accidents with replicating assemblers."

Indeed, a similar problem had confronted the genetic engineering people back in Cambridge. In 1976, Alfred Velucci, who was then the mayor of Cambridge, home of Harvard and MIT, led a fight to ban all recombinant DNA research inside the city limits. The people who worked in those labs were "Frankensteins," he said; they could, quite accidentally, create monsters and let them slip out of the laboratories and into the city, where they could attack people, disrupt the food chain, destroy downtown Boston.

"If worse comes to worst we could have a major disaster on our hands," Velucci said.

Since Drexler originally came up with the idea for assemblers, one of his biggest worries was that fear of a nanotechnological holocaust would lead, in the same way it did in the case of genetic engineering, to a ban on research, which he thought would be an unparalleled tragedy. We'd lose all the benefits that nanotechnology could have conferred on us, without necessarily escaping the dangers, because if *we* didn't develop it, someone else would. Whoever got nanotechnology first, in Drexler's view, could end up dominating the world.

Initially, Drexler was frightened by the gray-goo problem. "Between the time that I thought of the idea of assemblers and nanotechnology in early 1977, and when I began preparing my paper in 1980, I said almost nothing about these ideas because I was afraid of the possible consequences in terms of accidents or abuse. Later I learned that abuse was the real issue and that accidents were so easily avoided that they were a very secondary concern."

You could prevent an "accident," he learned, in any number of ways, by programming assemblers to cease functioning after a certain number of replications, for example. Or you could enclose all nanotechnology research inside of sealed laboratories, making the escape of dangerous replicators impossible. Or you could build assemblers that functioned only in the presence of a certain "vitamin," found only in the lab, and so on.

There was a whole list of things you to could do to confine gray goo, and in fact the more he considered it, the less of a problem he thought the gray goo was. The saving grace of assemblers was that they were just *machines*.

"For an industrial replicator designed to operate in a vat of fuel and raw material chemicals, for that to accidentally turn into a replicator that's able to survive in nature, well, that would be about as likely as a car—just 'by accident,' in the garage—being able to wean itself from its diet of gasoline and transmission fluid and go out and live on tree sap in the wild."

The *real* danger, he decided, lay in the deliberate abuse of replicators by foreign governments, terrorists, or even individuals who might have gotten their hands on some nanoweaponry. Conceivably, they could threaten the world with molecular devastation unless their demands were met, creating a situation as dangerous as global thermonuclear war.

The way out of this was not immediately apparent, but one of Drexler's ideas was to fight nanoweapons with nanodefenses. "We can build nanomachines that act somewhat like the white blood cells of the human immune system: devices that can fight not just bacteria and viruses, but dangerous replicators of all sorts." Drexler called such devices "active shields," because they'd be dynamic instead of fixed: they'd engage in various defensive tactics against different invaders.

So once again nanotechnology seemed omnipotent: threats posed by "bad" assemblers could be counteracted by "good" assemblers. After a while, even some hard-core cryonicists got tired of constantly hearing about nanotechnology's litany of miracles. Alcor's Thomas Donaldson, for example, compared the much-talked-about Nanotechnological Era to the Coming of the Apocalypse.

"I have noticed, too much, both in cryonics and out, a strong desire to interpret nanotechnology in the exact terms of Christian myth," Donaldson wrote in an issue of *Cryonics*. "It's as if a person carries out a renaming exercise: God = Nanotechnology, Drexler = Christ. (Sorry, Eric!)"

By the spring of 1988, nanotechnology had become mainstream enough so that Eric Drexler was invited to give the world's first college course on the subject at Stanford University. Drexler had already moved to California from the East Coast; California was, after all, where many of the advances necessary for nanotechnology were going to come from: computer miniaturization, microelectronics, artificial intelligence work, and so on. Anyway, when he got to California, Drexler was made a Visiting Scholar at Stanford University, and one day Nils Nilsson, who was chairman of the computer science department, had lunch with Drexler and asked him what Stanford ought to be doing about nanotechnology.

"Probably nothing formally," Drexler said. "I don't think the field is quite far enough along yet."

A couple of weeks later, though, Nilsson came back and asked Drexler whether he would teach a course on the subject, just one session a week. Well, that would be easy enough, Drexler thought, and besides, it would be an opportunity to work through some of the material for the new book he was writing, so he agreed.

It was too late to put the course in the Stanford catalog, or even in the class schedule for the upcoming semester, so they had to rely on posters and word of mouth. Nobody knew just what to expect as far as enrollment was concerned, so the course was scheduled for a small room meant for about thirty students. When class time came, some eighty students were packed into the room, sitting on the floor, standing, overflowing out into the hallway. It seemed that not another soul could possibly fit in there.

But then there was a latecomer. He was a bona fide, fully enrolled, fully registered student, and tonight of all nights he wasn't about to make do out in the hallway. So with a roomful of students, plus Mr. Nanotechnology himself looking on, he climbed in the window.

Back on the East Coast, Stanley Schmidt, editor of the science fiction magazine *Analog*, saw that nanotechnology could give birth to a whole new era of science fiction. He had read Drexler's *Engines of Creation* and devoted an editorial to it, "Great Oaks from Little Atoms."

"I'm not often willing to devote so much of an editorial to what amounts to a book review," he said, "but once in a while a book comes along which really needs to be read by anyone seriously interested in the future, and which belongs on the most accessible shelf of the working library of anyone who wants to write real science fiction. This is one of them."

A little more than a year later, *Analog* writers had gotten the message. A story called "The Gentle Seduction," by Marc Stiegler, depicted one character telling another how Drexler's marvels would revamp the solar system.

"With nanotechnology they'll build these tiny little machines—machines the size of molecules. They'll put a billion of them in a spaceship the size of a Coke can and shoot it off to an asteroid.

The Coke can will rebuild the asteroid into mansions and palaces. You'll have an asteroid all to yourself, if you want one."

But to be accurate, Stiegler's was not the very first science fiction story on nanotechnology. That distinction belongs to Robert Heinlein, the science fiction visionary who'd had himself cremated and his ashes scattered off the California coast. He'd written a story called "Waldo" and published it under the pseudonym Anson McDonald in the August 1942 issue of *Astounding Science Fiction* (*Analog*'s predecessor). This was well before Richard Feynman had ever talked about master and slave hands building "a billion tiny baby lathes," but Heinlein's story was about an inventor named Waldo who built gadgets that he named after himself, and which were therefore called "waldoes."

These gadgets consisted of robotic hands, the "primaries," which were worked by the human operator, and the "secondaries," which were the robotic hands themselves; originally the waldoes "had been designed to enable Waldo to operate a metal lathe."

The secondaries were smaller than the primaries, so that when the human operator worked the primaries, the secondaries did the same work on a smaller scale. Naturally, Waldo, the inventor, "used the tiny waldoes to create tinier ones."

Later, the smallest waldoes would be used to do tissue examination and repair. They had visual and sensory feedback so that the human operator could see and feel what he was doing.

"His final team of waldoes used for nerve and brain surgery varied in succeeding stages from mechanical hands nearly life-size down to these fairy digits which could manipulate things much too small for the eye to see. They were mounted in bank to work in the same locus. Waldo controlled them all from the same primaries; he could switch from one size to another without removing his gauntlets. The same change in circuits which brought another size of waldoes under control automatically accomplished the change in sweep of scanning to increase or decrease the magnification so that Waldo always saw before him in his stereo receiver a 'life-size' image of his other hands.

"Such surgery had never been seen before, but Waldo gave that aspect little thought; no one had told him that such surgery was unheard-of."

Soon after *Engines of Creation* came out some forty-five years later, though, not only was such molecular-scale nerve and brain surgery heard of, it was clear to cryonicists that nanotechnology would *have* to work if they were to come back and live again. It would *have* to be possible if their frozen heads were ever to be defrosted, revived, and fitted out with new bodies.

That was supposing you could even find them all. When, about a year after the Dora Kent crisis, her head had still not turned up anywhere, Curtis Henderson, who had cofounded the Cryonic Society of New York with Saul Kent, and who was, along with Bob Ettinger, one of the Truly Immortal Cryonics Poets, composed a little verse that summed up the whole situation quite nicely.

He adapted Sir Percy Blakeney's Scarlet Pimpernel rhyme and said:

> They seek it here,
> They seek it there,
> Those coroners seek it everywhere.
> Is it alive or is it dead,
> That damned, elusive frozen head?

5

Postbiological Man

*I*n 1972, about ten years after he'd written *The Prospect of Immortality,* Bob Ettinger was having some new thoughts about improving the lot of humanity. Obviously there was more that could be done for people than simply making them immortal; indeed, that was only a first step on the way to even greater heights. His new thoughts were directed toward extricating mankind from the tragic problem that commonly went under the heading "the human condition."

Supposedly, according to the Higher Philosophical Critics, "the human condition" constituted both the glory and the shame of the species. The glory was symbolized by all that was good and worthy about people: they possessed reason, creativity, feelings of empathy toward others, systems of ethics and religion, and so on. Mozart, Rembrandt, Shakespeare, all these were to the good. The shameful part was the way human beings had always botched things up, virtually since the dawn of time. Basically, mankind had an innate tendency toward war and violence, and more generally for letting civilization go to pot. The Inquisition, Hitler and the Holocaust, the plight of the homeless, decaying infrastructure, The Bomb— all these stood on the other side of the balance sheet. There was no end to the listing of human foibles, atrocities, and tragic flaws, so "the human condition" was commonly understood to be more bad than good, as was made clear by that roughly synonymous term, "the human predicament."

Anyway, after he got the cryonics show on the road Ettinger decided to take this "human condition" business seriously, for he himself saw the cause of the problem in crystal-clear terms. People had, as he thought, "cheap bodies, erratic emotions, and feeble mentalities." Their bodies were subject to disease, disability, aging, and death; their minds were battlegrounds of warring impulses, drives, and emotions; human memory and intelligence, such as they were, could be improved upon drastically.

"To be born human is an affliction," Ettinger thought. "It shouldn't happen to a dog."

It was no consolation that all of these shortcomings were quite understandable in evolutionary terms. Mankind, after all, was a product of nature, and nature worked not by intelligent planning and conscious design but by the worst kind of trial-and-error blundering: try this, try that, and see what worked out. Mostly, things *didn't* work out, as was clear from the fact that the over-whelming majority of the species that had ever evolved became extinct soon afterward. So it was no surprise that human beings were as botched up as they in fact were.

Others had made similar points in the past: they'd looked at man, seen the numerous flaws in the engineering, as it were, and made proposals for overcoming them. There was the case of David Hume, for example, the Scottish philosopher who noticed back in the 1700s that people would be much better off if only they'd been designed a little more intelligently. There was no inherent reason why they had to suffer *pain,* for example. "It seems plainly possible to carry on the business of life without any pain," he said. "Why then is any animal rendered susceptible of such a sensation? If animals can be free from it for an hour, they might enjoy a perpetual exemption from it."

By the last quarter of the twentieth century, science had advanced to the point where redesigning animals had not only become pos-sible, but had already been done. Indeed, some newly invented animal species—never before seen in nature, products exclusively of the laboratory—had been submitted to the United States Patent and Trademark Office for patents, and had received them. In the late 1980s two Harvard University inventors, Philip Leder and Timothy Stewart, built a better *mouse,* of all things. Theoretically, this new mouse would be valuable in cancer research, so in 1988 a patent on the animal was assigned to the Du Pont Company, which called the new species "OncoMouse."

Well, if you could design new animals, the next logical question was, why not new humans? After all, which of nature's creations stood more in need of improvement than man himself, as many forward-looking scientists, and even their young children, had realized.

Freeman Dyson told of the time his five-year-old adopted stepdaughter first saw him naked. "Did God really make you like that?" she asked him. "Couldn't he have made you better?"

Dyson regarded this as rather perceptive. "That is a question," he said afterward, "which every scientific humanist should be confronted with, at least once in a lifetime. The only honest answer is, of course, yes. I cannot regard humanity as the final goal of God's creation. Humanity looks to me like a magnificent beginning but not the last word."

Then, too, there was the utterly unanswerable Argument from Hitler to contend with, as in this version by Doyne Farmer, a researcher at the Los Alamos National Laboratory: "As a scientist I'm constantly frustrated by the inadequacies of my own brain. I'm frustrated by the inadequacies of people. I mean, any species that can let Adolf Hitler run a major geographic region for fifteen years is seriously flawed. I don't want to knock human beings too much, human beings are great. But why should we be restricted to human nature? Why shouldn't we go beyond?"

Now a person might offhand think that the so-called humanists, those whose business it was to study "the human condition," would have a little better perspective on what the Argument from Hitler really meant. Maybe there was some way of forgiving, or at least excusing, mankind from this moral lapse.

But no. Robert Nozick, Harvard University's star philosopher, once contemplated the meaning of Hitler and the Holocaust.

"It now would not be a *special* tragedy if humankind ended," he said. "Earlier, it would have constituted an *additional* tragedy, one beyond that to the individual people involved, if human history and the human species had ended, but now that history and that species have become stained, its loss would now be no *special* loss above and beyond the losses to the individuals involved. Humanity has lost its claim to continue."

In fact, some went even further than this, saying that it would be *a good thing* if the human species went the way of all flesh. "The death of *Homo sapiens* is an evil [beyond the death of the human individuals] only for a racist value system," said Frank Tipler, the physicist. "Our species is an intermediate step in the infinitely long temporal Chain of Being that comprises the whole of life in space-time. An essential step, but still only a step. In fact, it is a logically necessary consequence of eternal progress that our species become extinct!"

That was the news from the world of advanced theoretical physics. So anyway, if mankind wasn't the last word, if human nature was something that could and ought to be surpassed, if humanity had lost its claim to continue, and if in fact its extinction was necessary for eternal progress, well then it was high time to get on with the job. Time for a major overhaul, one that would take the human animal to a new level, to a more fitting, *trans*human condition. It was high time for this, because now, at last, we had the power and means for actually making the change.

No sooner did you start redesigning man, though, than you'd be charged with every known form of metaphysical felony: arrogance, hubris, "playing God," and all the rest. And of course the charges would be entirely accurate. Didn't Hume say that *even God* could have done a far better design job than he did?

"It would have been better to have created fewer animals," Hume had said, "and to have endowed these with more faculties for their happiness and preservation."

Didn't Bob Ettinger have some rather blunt criticisms to make of Mother Nature?

"It's hard to imagine that human engineers could be any clumsier or messier than that old slattern Dame Nature," he said. "The 'normal' processes of evolution are wasteful and cruel in stupefying degree. Dame Nature considers every species and every individual expendable, and has indeed expended them in horrifying numbers. Even an occasional calamitous error in planned development could scarcely match the slaughter, millennium in, millennium out, of fumble-fingered Nature."

Other foes of human engineering claimed that the very notion of an imperfect being somehow "perfecting itself" was inherently self-contradictory and impossible. But to Ettinger this was just misplaced pessimism. People had *always* tried to improve themselves, both mentally and physically, with everything from self-discipline and exercise to medicines, eyeglasses, and hearing aids, so there was no fundamental difficulty in the thought of a flawed species raising itself by its bootstraps.

"We can often do indirectly, and by stages," Ettinger said, "what at first seems quite beyond our scope."

In any event, Ettinger thought that you wouldn't have to *invent* a superman so much as *assemble* him from ingredients already on hand. We had real-life examples of rare intelligence and creativity, in people like Newton and Einstein, and beyond that there were plenty of characters in literature whose abilities we could emulate. Sherlock Holmes, for example, was a man of perceptiveness, imagination, and rare deductive powers. Why not build exactly such talents into our superman? There were examples from the machine world, too, where it was obvious that humans were, for the most part, quite outclassed. Almost without exception, machines did things far more efficiently, faster, and more cheaply than people did.

Designing a superman would be no real problem, not with all the examples of superhuman attributes and powers that we had in front of us. The only problem would be actually putting the plans into practice.

Bob Ettinger put all this into his new book, *Man into Superman,* where he explained everything in great detail. We could get beyond our shameful backwardness. We could rise up out of our primeval and barbaric state. We could reach the point where we could only look back and wonder why we had to suffer for as long as we did in that temporary, inglorious, and intermediate stage known to all as "the human condition."

At about midcentury, Arthur C. Clarke realized that his very first work of science fiction, the genre that was supposed to be way ahead of its time, was being threatened by the normal and ordinary progress of science. Just plain science was making some of his farthest-flung narrative projections obsolete. These were contained in his novel *Against the Fall of Night,* which had grown out of a short story that he originally began in 1935, at the age of eighteen. Rather ambitious, it told of a society in the extremely distant future, millions and millions of years hence, when people lived in peace and plenty although in constant fear of an unseen enemy. From the beginning though, Arthur Clarke had been dissatisfied with his story.

"It had most of the defects of a first novel," he admitted twenty years later, "and my initial dissatisfaction with it increased steadily over the years. Moreover, the progress of science during the two decades since the story was first conceived made many of the original ideas naive, and opened up vistas and possibilities quite unimagined when the book was originally planned. In particular, certain developments in information theory suggested revolutions in the human way of life even more profound than those which atomic energy is already introducing."

Some of the "developments in information theory" that he spoke of were contained in Claude Shannon's article, "The Mathematical Theory of Communication," which was published in 1948. Clarke met Shannon a few years later, in 1952, at Bell Laboratories, where Shannon worked. At the core of Shannon's piece was the insight that information of any type whatsoever could be encoded in the form of binary digits, or bits, and then communicated as a series of electrical impulses. This was important because of the generality involved: literally *any* information—a dictionary entry, the score of a symphony, a picture—could be reduced to controlled bursts of electrical energy. Separately, it had been known at least since the 1930s that there was electrical activity going on in the brain, so it was conceivable that memory, and perhaps even the human personality itself, existed in the form of electrical impulses. All at once it seemed that there was a deep link, never before noted, between man and machine.

Arthur Clarke wondered what might follow from this discovery. Whatever else it was, the brain was an information storage organ. If its information was stored in the form of electrical impulses, then it was possible that those same impulses could be detected by an electromechanical apparatus and reproduced in another medium, such as a memory bank.

The implications of this were startling. If you got enough information out of the brain and reproduced it with enough accuracy elsewhere, then you'd have a way of re-creating that person's memories, their innermost thoughts, feelings, and everything else. You'd be able to remake the person in a form other than his original flesh-and-blood physique. You might even be able to transfer an entire human mind into a computer. It was hard to know what this would *mean*, exactly, but at the very least it was clear that our concept of what it was to be a human being would be changed forever.

Arthur Clarke put all this into another novel, *The City and the Stars*, which was published in 1956. This one also took place millions of years in the future, but this new story at least *felt* like it: people had total control over their own minds and memories. They had learned, over the course of millennia, how to extract a mind from the brain. "We do not know how long the task took," one of the novel's characters explains. "A million years, perhaps—but what is that? In the end our ancestors learned how to analyze and store the information that would define any specific human being—and to use that information to re-create the original."

How they managed the trick was not explained, but the underlying concept was stated quite clearly. Everything hinged on the fact that the human personality was, in essence, *information*.

"The way in which information is stored is of no importance," the narrative said. "All that matters is the information itself. It may be in the form of written words on paper, of varying magnetic fields, or patterns of electric charge. Men have used all these methods of storage and many others. Suffice it to say that long ago they were able to store themselves—or, to be more precise, the disembodied patterns from which they could be called back into existence.

"A human being, like any other object, is defined by its structure—its pattern. The pattern of a man, and still more the pattern which specifies a man's mind, is incredibly complex. Yet nature was able to pack that pattern into a tiny cell, too small for the eye to see.

"What Nature can do," Arthur Clarke wrote, "Man can also do in his own way."

*A*rthur C. Clarke may have gotten there first—coming up with futuristic new ideas was, after all, his business—but plenty of others were latching on to the same notion only a few years later. Indeed, once word of the computer revolution got around, the notion of putting a human mind into a machine had several independent incarnations, spontaneously generating all over the place, as if it were an idea whose time had come. The difference was that now it was not meant as fiction, but as something that might actually be accomplished through science.

One of the first to take this seriously was Frederik Pohl, in his 1964 article "Intimations of Immortality," published in *Playboy*. Pohl normally wrote science fiction, but this time he was writing science *fact*, or at least he was claiming to. He talked about the different ways of prolonging human life: removing the causes of death; direct control of the aging process; Bob Ettinger's cryonics proposals. All these things could be done to preserve the *body*. Nevertheless, "the essential 'you' isn't your body," he said. "It is what we will call your personality, your memory, or your mind." This essential *you* could be preserved inside a computer, "a collection of magnetic impulses in an IBM machine."

As far-out as it was, Pohl's idea was not quite as advanced as the one that had been offered by Arthur Clarke some ten years earlier. Instead of reading out the contents of your mind and transferring it into a memory bank, which was Clarke's method, what Pohl imagined was that you'd take a computer and start educating it, bringing it up as though it were a child.

"We read it *Moby Dick* and *Treasure Island* and we teach it the words of *Nuts to the Bastard King of England* and *Gaudeamus Igitur*. We teach it the flavor of a vodka gimlet and the scent of the back of a pretty girl's neck, the feel of a clutch in a Stingray and the sounds of Mozart and Monk. We teach it, in short, everything you know."

Somehow, through feats of dextrous programming, all these unrelated bits and pieces were supposed to coalesce inside the computer until "you, or something like you," sprang into operational life. Whatever it was that came to life in there, it would be immortal ever afterward—or at least for as long as the computer kept running.

Here was hard science already creeping up on Arthur Clarke's revised futuristic projections only a few years after he'd advanced them. Still, Pohl's scenario had the disadvantage that the computerized person would no longer be *himself* in any important way. He'd only be somewhat *like* himself, an approximation. The reason for this was that Pohl saw no way of actually doing the type of direct mind-to-metal transfer that Arthur Clarke had written about. All this stuff about "reading out" the contents of a human brain was well and good, but until someone came along with a specific, realistic method for actually doing it, it would have to remain in the realm of fantasy.

The truth of this realization was reinforced a few years later when an IBM employee by the name of Dick Fredericksen also began to think about placing human minds into computers. Fredericksen, who had a master's degree in information science from the University of Chicago, was then a researcher at the IBM Thomas J. Watson Research Center, at Yorktown Heights, New York. For a while he published his own newsletter, a mimeographed bunch of reflections on diverse subjects, called *A Word in Edgewise*. He'd send copies to his friends, relatives, and other interested parties, who would pass them around, post them up on bulletin boards, and so on. Anyway, it was a tragic day for Dick Fredericksen when he got a letter from home telling him that his sister Kaye was ill with Wilson's disease, a blood disorder that had allowed traces of dietary copper to accumulate in her body to the point where she was slowly being poisoned to death.

This was happening just at the time when heart transplants were in the news, so Fredericksen wondered if it ever might be possible to do a liver transplant, for Kaye's liver was about finished. And then, a bit later, as her nervous system was attacked too, he wondered whether her entire body could be replaced by another one, a "body transplant."

But then he went even further, asking himself whether you could forget about bodies entirely and, as he put it, "implement the human being in alternative hardware," specifically, the computer. Fredericksen became so intrigued with the concept that he ended up writing an eighty-page description of the idea and running it in four successive issues of his newsletter in 1971 under the title "I Have a Pipedream."

"Maybe we can read out the one into the other," he wrote. "Maybe a transplant is possible—memory, consciousness, 'soul,' and all. Having prepared an alternative vehicle for sentience, maybe we can climb into it. Maybe, in short, death is an unnecessary affliction."

Unfortunately, like Fred Pohl before him, Dick Fredericksen ran up against a blank wall when it came to the reading-out problem. "This is perhaps the weakest link in the whole chain," he said. "What does it *mean* to 'read out' the personality of a man from flesh to robot, in such a way that he actually experiences transplantation and survival? We are so far from having an understanding of this, that we haven't as yet read out a single message that we could interpret." Nevertheless, it was only a matter of time, he thought, before scientists learned how to manage the task.

Later, even Bob Truax got into the act. He was writing a book, *The Conquest of Death,* in which he proposed seven different methods for becoming immortal. Most of them involved getting rid of the present human body, which, from Truax's own engineering perspective, was riddled with defects.

"What right-minded engineer," he asked, "would try to build any machine out of lime and jelly? Bone and protoplasm are extremely poor structural materials." (Arthur Clarke, separately, had made the same point with regard to the human eye: "Suppose *you* were given the problem of designing a camera—for that, of course, is what the eye is—which *has to be constructed entirely of water and jelly,* without using a scrap of glass, metal, or plastic. Obviously, it can't be done.")

So it would be no great loss, Truax thought, if you *got rid of* the human body and replaced it with something else that was stronger, better designed, and more suitable to an extremely long life span. In fact, it might not even be such a bad idea to dispense with the notion of *bodies* altogether, for even a backyard rocket engineer like Bob Truax could see that the core of the human personality was not matter but mind: "It has been called the 'soul,' the 'id,' or simply the 'self' or 'identity.' Certainly it is not the body."

Truax thought that the essence of the human personality was in fact the *memory*—"If I can remember who I am, then I continue to exist," he said—so why not take a human mind, transfer it to a computer, and let that person's memories come alive inside it? The advantage of this arrangement was that once you had the person stored away in the computer, then you could create something that human beings had never before had, *backup copies*.

"One would keep a copy of the mental program on file in a vault, or several copies in several vaults, so that when and if the original is destroyed, the program is simply copied into the latest model of 'genus homo.'"

But of course all of these advanced schemes assumed that the reading-out problem could be solved, and although he was sure that eventually it would be, not even Bob Truax could offer any realistic proposals as to how it could in fact be done. So all of these fine thinkers and visionaries, Arthur Clarke, Fred Pohl, Dick Fredericksen, Bob Truax, and God knows how many others, all them were having these wonderful hubristic dreams of putting themselves away into memory banks, making backup copies, and living forever, but not one of them could figure out a way of getting to first base insofar as actually *doing it* was concerned. They couldn't read out their own mental data. It was almost as if the mind were an occult entity, hidden away inside the brain, where it was impossible to grab hold of by any direct action.

That was the situation, anyway, as Hans Moravec came on the scene.

*H*ans Moravec was born in Kautzen, Austria, in 1948 and emigrated with his parents to Canada four years later. From the time he was a kid he amused himself by making toy machines. It all started with a game set, called Matador, that he was given at the age of three. Basically Matador was a set of wooden blocks, pegs, wheels, pulleys, and such, all of which could be fitted together in different combinations to build just about anything you wanted.

Hans's father was an electronics technician and helped him put different gadgets together: they made wagons, a toy car, a hand-cranked machine that hammered nails. And then they made a dancing man.

The dancing man was different from those other things: it moved as if it were somehow "alive." It was nothing more than a couple of blocks held together by wooden pegs (they formed its head and body), plus wooden slats for arms and legs. The whole thing rested on a box that had a crank at the side; when you turned the crank this mechanical man would bounce up and down, jittering and bobbing and dancing around as if it had a life of its own. It was an extremely primitive device, but that didn't matter: the thing *moved,* which was the important part. "That's not a man," Hans thought to himself at the time, "it's just some blocks. But it *acts* like a man."

That was only the beginning of Hans Moravec's obsession with robots. Later, in Canada, when he was in the fifth grade, he read an article in *Junior Messenger* magazine about a girl who had built a robot, and there were even pictures of it. You could see that it was shaped like a person, but its insides were made out of electrical wires and switches and it had this little light bulb in its chest that blinked on and off. That was its beating heart.

Well! This was *quite* an advance over his old dancing man, and so naturally Moravec had to build one of his own. Fortunately his father had an endless supply of electrical equipment down in the basement, and Hans used some of it plus motors taken out of toys he'd disassembled, and then he put together his first real robot. The body was just a tomato-juice can, but after Hans put a toy motor inside and had connected it up to a gear train that reached to the arms, the thing took on an entirely new aspect: it too seemed, in its way, "alive." The motor ran off an internal battery, and with the current switched on, the arms worked back and forth under their own power, without your having to turn any cranks from the outside. This little self-contained live wire—"Tin Man," as he called it—was an important milestone in the life of Hans Moravec.

At some point in his growing up Hans got into his head the tiny schoolboy conceit that *he himself might be a robot.* He didn't believe it in any *real* sense, nothing serious, it was just a fun thing to imagine now and again. And besides, it was just barely possible. At least it wasn't *im*possible. "Well, what if *I'm* a robot? *Ha, ha, ha!*"

Later on Moravec would read about Truly Advanced Robots in the pages of science fiction. One of his favorite stories was a novel by A. E. Van Vogt, *The World of Null-A,* in which there was a machine that turned out to be much smarter than the people. "Here were these people pathetically trying to think straight and this *machine* was doing it in spades and a thousand times faster!" said Moravec. That was in fiction, of course, but Moravec was convinced that the same was true in real life, that machines were generally superior to people.

"You don't have to look far to see that earth-moving machinery can dig better than people, airplanes fly better, boats go through the water faster. It's a very small extension to imagine that it's possible to make a computer that can think better than a person."

It was in high school that Moravec first got the idea for what he later called "downloading," transferring the contents of a mind into a computer. He and a friend by the name of Ken Simonelis got to arguing about whether intelligent robots would actually *be* people or would only be *like* people. Moravec, who by this time sincerely *wanted* to be a robot, thought they'd actually *be* people, a view, incidentally, that he later renounced as demeaning (to the robots). Intelligent robots, he decided, would be superior to people on any possible criterion. They'd be *far* more intelligent, talented, and powerful than human beings ever were or ever could be, unless and until humans evolved into something better. But Ken Simonelis took the more conservative position, arguing that no matter how close robots came to being *like* people—they could look like them, act like them, even "think" like them—nevertheless, they *weren't* people, they were only robots, "machines."

For a long time there seemed to be no way of settling the argument one way or the other, but then one day Moravec had an inspiration.

"I thought of a method by which I could convince him. Assuming that a human being is fully explained by the physical interactions of his parts—which he accepted—I said, 'Look, suppose you took a human being and started replacing his natural parts with equivalently functional artificial parts, and you did this on a very small scale, neuron by neuron, or whatever. At the end, what you'd have would be something that still *worked* the same, because by definition each individual part worked just like the part it replaced, only it was made of something else: metal, or plastic, or whatever. But what you'd have at the end would still be a human being.'"

Indeed it was hard to see anything wrong with this reasoning. If a person with a wooden leg was still human, then so was a person with two wooden legs, and so on. Once you'd started down this slippery slope there seemed to be no place where you ever had to stop—where was there a dividing line?—so Moravec claimed that even a person with a *wholly* nonbiological body would still be a human being.

But that was only point one. The next was to imagine *making* an artificial human being, from scratch. In other words, instead of starting out with a normal, biological human and going through that tedious part-by-part replacement business, you'd begin by building the artificial human *directly,* from out of the parts bin. You'd put the parts together in the right order until what you had standing there in front of you was a Man-made Man.

Then for the third and final step. What you'd do is, you'd give this artificial human the *mind* of an ordinary human being. You'd take a normal adult and read out his or her mental store of information, thoughts, and memories, and transfer it all bit by bit into the artificial human's head.

"This thing," Moravec said, "could now carry on the life of the person whose mind you transferred to it. The robot would have all the same skills and all the same motivations as the human being did, and so it could raise the children or do anything else the human could do. In fact, and for all practical purposes, this 'robot' *is* the human being. It does the job of the human being—it quacks like a duck and acts like a duck—it interacts with its friends just like before. *Everything* the human being did, this artificial replacement does too. So if you don't want to call it a human being, it seems like just perversity on your part."

Ken Simonelis could never go this far, but many others did, and in fact when Moravec arrived in 1971 at SAIL, the Stanford Artificial Intelligence Laboratory, what should he find tacked up on the bulletin board but a copy of Dick Fredericksen's "I Have a Pipedream." Fredericksen too had worked it all out, although from a different starting point.

The downloading stuff, admittedly, was not universally accepted even by all the people at SAIL. About half of them took it seriously and half of them didn't, which is where it remained ever afterward. But none of Moravec's colleagues thought that the mind-transferral business was *unbearably* flaky, and at any rate it never held Moravec back or stopped him from rising up through the ranks. By the time he was director of the Mobile Robot Lab at Carnegie-Mellon University, in Pittsburgh, he had developed the scenario to the point where he could put it in print and not have people die of heart attacks when they read it.

In fact he wrote a whole book about it, called *Mind Children.* At first he bombarded the country with the book in manuscript form, circulating it around to all his AI friends and comrades— "the artificial intelligentsia," in Joe Weizenbaum's phrase—to get their critical advice and suggestions. The book went through three main drafts, and finally, in the fall of 1988, the finished product was published by the most prestigious academic press in the country, Harvard University Press.

It explained in detail how *people could become robots,* how they could *download themselves into computers,* and how all of this could be done *in the next fifty years.* Most important, Moravec had solved the reading-out problem, coming up with not just one but *four* different methods of getting the mind out of the brain and into the computer: "transmigration," he called it (a bit tongue in cheek), as in "transmigration of souls." Naturally, his own preferred method was to have the whole task managed by robots, the super-intelligent robotic surgeons of the future.

Moravec gave the scene in *Mind Children: "You've just been wheeled into the operating room. A robot brain surgeon is in attendance. By your side is a computer waiting to become a human equivalent, lacking only a program to run."*

The actual procedure begins with you, the patient, fully conscious, with only your skull anesthetized. The robot surgeon opens your cranium, places his mechanical hands on the brain's outer surface, and starts taking data from the first layer of brain cells.

The surgeon has all manner of advanced machinery at his disposal, and with it he writes a program that simulates the functions of the brain cells he's just scanned. He enters that program into the computer nearby—the one that's going to become *you*—then runs the program so that you can experience its output while you're still lying there on the operating table.

Suppose, for example, that these first brain cells contained, among other things, your mental picture of Einstein—an image of the frizzy hair, the drooping eyes, the sad expression, and so forth. When the surgeon ran the program that contained your Einstein image, you'd be able to see the mental picture the program produced, and you'd be able to compare it against the original mental picture that was still there in your biological brain. If the two pictures didn't match up, the surgeon would adjust the computer code until it reproduced your original mental picture faultlessly.

Once the downloaded program had produced the correct picture, the brain cells that had originally contained it would have become superfluous, so out they'd go, never to be seen again. (And good riddance! As compared to advanced computer hardware, Moravec thought, brain cells were slow and unreliable. Besides which, damaged brain and nerve cells didn't even regenerate themselves. All in all, brain cells were not one of Mother Nature's master achievements.)

Then the surgeon repeats the whole process on the next innermost layer of cells, sensing the electrical activity going on in and among them, adding that information to the cumulative program now running inside the computer, checking it against your conscious experience and making the necessary corrections. When there's a perfect match, this new layer of cells is stripped away too, and the process continues.

What was this but an advanced version of the part-by-part replacement strategy that Moravec had been thinking about since high school? You're replacing a biological part with a computer-chip part, but so what? Since the two media have identical output, the mind in the computer is arguably human.

Maybe at the halfway point—when you're half in the computer and half still there in your outmoded body—there'd be a celebration of some sort. Maybe you'd drink a glass of champagne, and the surgeon—no robot drinks for him, please!—would propose a toast. Or something; these details are hard to foresee.

The process would continue until there was nothing left of your biological brain at all. Your brain pan would now be empty, looking much like an ashtray, but *you*—your mind, your memories, your entire identity, personality, and conscious experience—all of *these* would now be inside the computer.

"In a final, disorienting step the surgeon lifts out his hand. Your suddenly abandoned body goes into spasms and dies. For a moment you experience only quiet and dark. Then, once again, you can open your eyes. Your perspective has shifted. The computer simulation has been disconnected from the cable leading to the surgeon's hand and reconnected to a shiny new body of the style, color, and material of your choice. Your metamorphosis is complete."

*B*ook reviewers read this and the other mind-boggling claims in *Mind Children* and some of them went berserk. Indeed, the prospect of a human being . . . *becoming a computer program!* . . . well, it was hard to swallow, to say the least.

Martin Gardner, the former Mathematical Games columnist for *Scientific American,* compared the downloading scenario to what he called "the Tin Woodman Conjecture." This referred to an incident in *The Wonderful Wizard of Oz,* the novel by L. Frank Baum, where the woodman, who had started off as an ordinary flesh-and-blood person, was changed into a tin man by the very part-by-part replacement method that was now being advocated by Hans Moravec. Parts of the woodman's body were chopped off and metal equivalents swapped in until he was a tin man through and through. There were other such stories in science fiction, but the difference between *them* and *Moravec,* Martin Gardner said, was that at least the fiction writers (thank God) "did not take their scenarios seriously."

Moravec could not see that this was much of a criticism. He'd been reading exactly such science fiction tales all his life—about mechanical men, intelligent machines, and so on—and *he,* for one, had always thought the stories were quite realistic. The only thing that was incomprehensible to *him* was when these mechanical men were portrayed as harboring secret desires to become "people." *That* was unbelievable.

"There was a 'Twilight Zone' episode in which Robert Culp played a character who discovered that he was a robot," Moravec said. "He hadn't known this before, he'd always thought he was a person, but then he finds out the truth—he's really a robot—and when he discovers this he's horrified. But my feeling was, *What the heck are you complaining about?*

"I didn't understand it at all. I never understood why Pinocchio wanted to become a real boy. And then Isaac Asimov had a story, 'The Bicentennial Man,' in which one of his humanoid robots wants to become a person. That's a cute story, but I read it and I thought, *Why in hell do you want to become a man when you're something better to begin with?* It's like a human being wanting to become an *ape!* 'Gee, I really wish I had more hair, that I stooped more, smelled worse, lived a shorter life span.'"

Which is not to say that Moravec looked down upon the animals. Just the reverse: he wanted to download *their* minds too, so that he could combine their special talents with his own.

"I assume that animals are not the nonentities that human beings sometimes treat them as. The long evolution that, say, a bird has behind it has produced some skills that are not found anywhere in the human race at the moment. These skills will be of interest—the ability to localize sounds, the ability to perch, echolocation, flight, all of those things. You might well like to borrow those skills as much as any person's carpentry skills, for example."

Bob Truax had once thought of a similar possibility, the process of combining two minds into one, giving you at the end what he called an "expanded person."

"Visualize a sort of super 'bar mitzvah,'" Truax had written in *The Conquest of Death*, "where the offspring, suitably wired for thought transfer, receives the knowledge, the memories of the father—or the mother, for that matter. If the process were continued for many generations, a great intellect, having roots in the dim past, could well result."

Moravec now said the same thing in *Mind Children*, describing how you could selectively merge with many other minds. You could "remember" *other* people's thoughts, you could experience the things that *they* had done, naturally losing much of your own identity in the process. But so what?

"In the long run you will remember mostly other people's experiences, while memories you originated will be incorporated into other minds. Concepts of life, death, and identity will lose their present meaning as your mental fragments and those of others are combined, shuffled, and recombined into temporary associations, sometimes large, sometimes small."

The final result would be a world-scale consciousness. "Our speculation ends in a supercivilization, the synthesis of all solar-system life, constantly improving and extending itself, spreading outward from the sun, converting nonlife into mind."

Flabbergasted reviewers described *Mind Children* as "a frightening tale," spoke of "the horror" of the postbiological world it conjured up, and classified Moravec as "yet another mad scientist." Moravec was so delighted by this outpouring of what he regarded as *human chauvinism* that he pasted up on his office wall some of their choicest comments:

> "The most lurid book ever published by Harvard University Press."—Noel Perrin, *Washington Post.*
> "Uncritical gee-whiz."—Mitchell Waldrop, *New York Times.*
> "Bizarre."—Martin Gardner, Raleigh, N.C., *News and Observer.*

Thankfully there were others—Fred Pohl, for example—who took a more enlightened view of downloading. Pohl, who had advanced his own proto-version of the process in his *Playboy* article, was particulary impressed by another one of Moravec's four downloading methods, the one that involved reading out the mind through the corpus callosum, a bundle of nerve fibers that sits midway between the two brain halves. Moravec had written:

"Suppose in the future, when the function of the brain is sufficiently understood, your corpus callosum is severed and cables leading to an external computer are connected to the severed ends. The computer is programmed at first to pass the traffic between the two hemispheres and to eavesdrop on it. From what it learns by eavesdropping, it constructs a model of your own mental activities. Ultimately your brain would die, and your mind would find itself entirely in the computer."

To Fred Pohl, this was an entirely reasonable prospect.

"What Moravec has done in his book," he said, "is to come up with an at least plausible-sounding technique for getting the mind into the machine—i.e., by splitting the corpus callosum and monitoring the impulses that pass between the two halves of the brain. I don't know whether that would really work or not, but at least one can see that ideally it might . . . which is more than anyone else has been able to suggest, to my knowledge."

But Moravec had other, even less messy ways of transferring the mind to the machine. You could wear a little portable computer around, like a Walkman. It would monitor your every move, record your every word and brain wave, and at length the machine would know you as well as you ever knew yourself. All this knowledge would then be put into a program.

"When you die," Moravec said, "this program is installed in a mechanical body that then smoothly and seamlessly takes over your life and responsibilities."

Or if even *that* sounded too much like work, there was the One Fell Swoop method—for those in a hurry to get out of their heads.

"A high-resolution brain scan could, in one fell swoop and without surgery, make a new you *While-U-Wait*."

So there you are inside the machine, inside this advanced computer that can hold the entire contents of your mind and memory. It even *feels* like you, so you're convinced that the mental transfer has actually taken place, that it's *worked,* that it's really *you* in there, just the same person you ever were—except that now you're no longer connected up to your body, which has already been disposed of.

But on the other hand you're wondering if this won't create a little problem, this business of being *confined to the inside of a computer*. What in the world are you going to *do*? How are you going to *have sex*? What are you going to *eat*?

The fact is, though, that essentially nothing has changed. Beforehand you were always confined to the inside of your *brain* whereas now you're inside the computer, so . . . what's the big difference?

The theory was, even though you're inside a computer you could still have *exactly* the same experiences you'd been having back in your old body. The only difference would be, now you'd be experiencing a *simulation* of reality rather than reality itself. Or perhaps it would be more accurate to say that you'd be experiencing a *different kind of simulation* of reality, because as far as Moravec was concerned, that's all your original body ever gave you: a simulation, a mental construct that the brain put together out of the data conveyed to it by the senses.

In fact, even Arthur Clarke had argued this way, back when he wrote *The City and the Stars*. The people stored in data banks could be made to have all kinds of synthetic experiences, but the fact that the experiences were *artificial* didn't seem to bother them at all. "Whether they were 'real' or not was a problem that had bothered few men for the last billion years," the narrative said. "Certainly they were no less real than that other impostor, solid matter."

And how utterly true this was. After all, didn't you, back in your old human body, only experience the world *indirectly*, via the intermediate agents of light rays, the retina's photoreceptors, the electrical impulses traveling up the optic nerve to the brain, and so on? At the *end*, it was true, those impulses were transformed into your "visual picture" of some external object or other, but the exact same thing could be done inside the computer. *Experientially* there would be no difference.

For Moravec, being inside the computer would be like seeing the same reality through a different pair of eyes, only *better* eyes than you'd ever had before, back when they were still those jelly-bean-camera things, handicapped by unsightly blind spots, whose focus always had to be corrected by means of spectacles or contact lenses, and so on. The eyes of the future would be far more versatile. For one thing, they wouldn't have to be imprisoned in your body.

"The best analogy right now is probably telepresence," Moravec said. "Suppose you put on this nice helmet, which has really great TV screens for your eyes to look at, and suppose that there's a robot somewhere else in the room, and it's got TV cameras for eyes. Say that its eyes are connected up to these TV screens in your helmet, so that you see what it sees. When you turn your head, the robot turns its head, and you instantly have the illusion that your center of consciousness is right there, two inches behind the robot's cameras."

Telepresence had been a big theme of artificial intelligence research ever since NASA latched on to it as a way of exploring the planets. You'd send a robot to Mars, or wherever, and it would collect all the sensations that a person could ever have up there and transmit them back to earth. Telepresence went by other names as well—"artificial experience," "virtual reality," and the like—but the main idea was that of projecting your consciousness out to wherever the mechanical sense organs were, and going there by *proxy*. (Moravec thought of it as "armchair exploration of the universe.")

Not that this would take any effort on the viewer's part. On the contrary, once you were hooked up to the robot's cameras, it would be an alteration of perspective that you couldn't avoid.

"It's a completely vivid and realistic experience," Moravec said. "You put this helmet on and you look around, and where are you? Of course you are where you're looking around from, which is the robot's head. So you're suddenly teleported into the head of the robot, and your sensation of consciousness is now over there. A downloaded consciousness will be something like that."

Grant Fjermedal, the science writer, once had such an out-of-body robotic experience. He'd visited a Japanese robotics lab, where he put on a helmet that was connected up to a robot fitted out with TV cameras for eyes. The robot was in another corner of the room, and as soon as Fjermedal began to look through the robot's eyes he felt that he was *over there*, over where the robot was.

And *then*, "Someone in the laboratory went over to the robot-mounted cameras and swung them around so that they focused on me. The walls spun during the maneuver, and then when the motion stopped and I was looking at myself, the out-of-body experience began. It was as if I were standing a few feet away *in another body* looking at myself. I moved my head to look up and down and even to look away. And when I looked away from that person who was me, it was as if that body were just another passerby."

Supposing that all this downloading, computer-consciousness, artificial-experience stuff was even remotely possible in a theoretical sense, the question that anyone in their right mind had to ask was: *Why do it?* Why bother going through all the bit-by-bit transfer and simulation, putting yourself into a computer, killing off your old body, then hitching yourself up to a bunch of new mechanical sense organs only to perceive the same old reality? Moravec's answer was:

"Mainly because of the travel possibilities. If you're now a piece of software that can be run in any available computer, then it's simply a matter of transferring that program into a waiting computer at the other end of the communication link."

You could be sent from New York to San Francisco in a matter of seconds, over telephone wires. Better still, when the next earthquake arrived you could get yourself *out* of San Francisco in a matter of seconds.

"You could fax yourself to a distant planet, star, or galaxy," Moravec said. "Perhaps you'd undergo some experiences at the remote location, and then, if you'd like, you can take your state of mind after those experiences, beam it back, and incorporate it with whatever experiences you've been undergoing at this end."

As bizarre as this stuff may have been, all of it followed straightforwardly from Claude Shannon's information theory, the theory that had inspired Arthur Clarke. "The way in which information is stored is of no importance," Clarke had written. "All that matters is the information itself." If that was true, then once people had been reduced to patterns of information, they could be broadcast to wherever you had the physical power to send them.

These electronic traveling methods might, of course, bring some new problems along with them. Say you're caught in San Francisco during an earthquake and you want to get yourself out of there in a hurry. But there are millions of other downloaded minds out there too, whereas (no surprise) there aren't enough telephone lines to go around.

But finally you manage to get a line out somehow, and *then* what happens? Then the line's a little . . . *crackly*, a little . . . *noisy,* there's so much . . . *static* on the line, because of all that rumbling and tumbling coming from the San Andreas fault or whatever it is, that when you get to the other end of the communication link you wind up as some . . . *half-witted bimbo!*

But in fact such a line of reasoning played right into Moravec's hands, for it gave him the opportunity to download to his listeners what was probably the single greatest benefit that would be conferred upon the human race—or, more properly, its successors— by his entire scheme: the possibility of *backup copies.*

And why not? Once you'd gone to all the trouble of having your mind read out of your brain, it would be stupid to settle for just *one* electronic copy of yourself. Anyone who'd ever used a computer knew from experience that no sooner did you get your most valuable data fixed in short-term memory than the power went off or the machine crashed or you touched the wrong keyboard button and all your hard work went up to electron heaven. So the first thing any smart computer user did was make *plenty of backup copies.*

Same thing with downloaded people: you'd make multiple copies of yourself and store them all over the place. That way, if one copy met up with some unpleasantness, another one could be activated immediately. You'd have to keep your copies up to date, of course, otherwise there'd be some temporary amnesia to get around, but by the time all of this became possible (which, according to Moravec, was in *just fifty years*) there'd be all kinds of software available for making backups automatically, day by day, hour by hour, or as often as you wanted.

All of which was an object lesson in just how risky human life actually was in an age when *not one person* had so much as *a single backup copy* of himself stored anywhere in the known universe.

Lack of backups—that was another unwanted feature of "the human condition," another one of nature's gross oversights that the species would shortly be able to leave behind.

Bob Ettinger's superman was basically an idealized human being. First of all, he wanted to get rid of mankind's more disagreeable or offensive qualities, so of course "the elimination of elimination" was high on the agenda.

"If cleanliness is next to godliness," he said, "then a superman must be cleaner than a man. In the future, our plumbing (of the thawed as well as the newborn) will be more hygienic and seemly. Those who choose to will consume only zero-residue foods, with excess water all evaporating via the pores. Alternatively, modified organs may occasionally expel small, dry, compact residues."

The toilet, as we know it, would be a thing of the past. Sexuality, on the other hand, would persist throughout the ages, with the transhuman's sexual capacity increasing in variety, intensity, duration, and just about every other imaginable way. New organs, new sexes, all these were in the cards. Nevertheless, Ettinger saw sexuality lessening in importance. "Eventually it will indeed become a smaller aspect of life than it now seems—not because sex will shrink, but because life will expand."

The transhuman body will be made more resistant to extremes of temperature, pressure, and so on, it being a major scandal that the original edition of the human body was suited for only a small portion of the very planet it evolved on. The earth is four-fifths water, but can human beings live in the oceans? *No.* And the earth is covered with a deep blanket of air, but can people fly through the air like the birds and the bees? *No.* So what was all this fulsome rhetoric about the "superbly adapted human body"?

And that was only looking at it from an earthly viewpoint. From a grander, more cosmic perspective, the human body was *far* less than what it could be. Humans were supposed to be a space-faring species (Ettinger thought), but look at what we had to bring up there with us: spacecraft, radiation shielding, air conditioning, heating, food, water supplies, radios, insulation, space suits. So when Bob Ettinger was designing his superman it was with an eye to overcoming all these obvious shortcomings, and more.

But the bodily refinements that Ettinger listed in *Man into Superman* were nothing compared to what Hans Moravec later offered in *Mind Children.* Moravec also wanted people to become supermen. In fact he himself had wanted to *be* Superman, the comic-strip character, ever since he'd first read about him.

"I read Superman comics when I was quite young—eight or so—and in the fifth grade we had to write an essay on what we wanted to become, and I wrote down that I wanted to become a reporter. Of course I didn't really want to become a *reporter,* I really wanted to be Superman, but I couldn't put that down—it just wouldn't fly. So I said I wanted to be what Clark Kent was, a reporter.

"But it turns out that I didn't really want to be Superman either. I really wanted to be Superman's archenemy, the scientist Lex Luthor. Superman was actually a *horrible underachiever!* Look what he was endowed with: X-ray vision, the ability to scan books in seconds, and so on, but here's Lex Luthor, a normal human being—who's even *bald!*—and he's able to come within a hair of outsmarting Superman, with just his unaided brain! Superman was all brawn. Lex Luthor was really the smarter of the two."

But then when he got the idea of replacing himself bit by bit with a computer simulation, Moravec realized that he actually *could* become Superman. "In fact this was even *better* than becoming Superman! You can kill Superman with Kryptonite, but with bit-by-bit transfer you can make *copies* of yourself! That seemed much better to me than anything I was reading about in science fiction."

Even Bob Ettinger, who spent lots of time redesigning the human body, knew that there was something better than being a glorified human. The possibilities were much greater if only the mind could somehow be coupled to a *machine.*

"In principle," he said, "a machine can be made to do anything that is physically possible, and if we envision a human brain coupled to a machine or complex of machines—so that the machines are extensions of the person—then, with only modest reservations, *we* can do anything, which means we can be anything."

Ettinger, though, didn't work much of this out; Moravec, by contrast, combined his bit-by-bit downloading invention with the idea of the most advanced, dextrous, and powerful type of robot he could think of, thereby coming up with his *adult* version of a superman, a *true* superman. As a matter of fact it would be nothing like a human being at all. A true superman, he now thought, would be a *Bush Robot.*

A bush robot (or robot bush; Moravec couldn't decide on the term), was the very last word in muscles and sensors; it possessed an almost infinite number of arms, legs, and other flexible links, each of which ended in photoreceptors far more sensitive than those ever seen before on earth. The "bush" aspect of it referred to the fact that each of the robot's limbs would branch out into smaller and finer limbs, like the twigs of a tree. The human body already did this in some small measure, for our arms end in hands that branch out into fingers. Such an arrangement was fine as far as it went, but for Moravec, nature had terminated the branching process too soon.

"There are many things the hands can't do. They can't hold seven things at once and are limited in the fineness of their manipulations. But all of this can be solved if you just carry the idea of a trunk, limbs, and fingers a lot further. If your fingers had fingers, and if those fingers themselves had fingers, and so on, then ultimately you could hold billions of things at once."

With his bush robot, Moravec took the idea of branching to the furthermost possible degree. The robot's arms would separate out and subdivide into billions of tiny extremities, ending up in a fanwork of several billion hairlike jointed structures, some of which could conceivably be small enough to manipulate individual molecules, *even individual atoms,* of matter. These billions of arms and legs could each move over an infinite range in all three dimensions, at rates of a million times a second, or more. They could rotate around on their own axis and expand or contract, like the shaft of a telescope. Each of the arms would be equipped with sensors that responded not only to light and heat, but to the full spectrum of electromagnetic effects.

More than that, some portions of the robot could separate off from the main robot, turning into smaller bushes. The smallest of these branchlets would be able to float through the air, like dust motes. Bigger ones, perhaps the size of insects, could walk on ceilings, like flies, or burrow into the earth, like worms. A bush robot would have all the capabilities, but none of the limitations, of all earthly animals rolled into one—not that it would look like any of them. It was a protean object, the very Platonic form of dexterity.

"A bush robot would be a marvel of surrealism to behold," Moravec said. "Despite its structural resemblance to many living things, it would be unlike anything yet seen on earth."

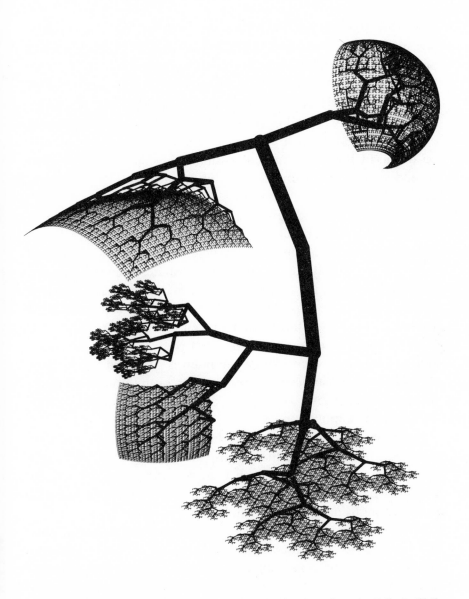

Nevertheless, there was a picture of a bush robot in *Mind Children*, an image composed of a quarter of a million line segments. It had taken a Macintosh II computer ten hours of running time to produce what was, to some eyes, anyway, a rather ghastly object. (Moravec himself, however, spoke of its "perpetual gracefulness.")

To control its billions of eyes and arms, the bush robot would be equipped with a correspondingly superintelligent, almost godlike brain. Put the two together and you'd have an almost omnipotent being on your hands. There'd be virtually no task, mental or physical, that it would be unable to accomplish.

"A trillion-limbed device, with a brain to match, is an entirely different order of being," Moravec wrote. "Add to this the ability to fragment into a cloud of coordinated tiny fliers, and the laws of physics will seem to melt in the face of intention and will. As with no magician that ever was, impossible things will simply *happen* around a robot bush. Imagine inhabiting such a body."

*T*he whole downloading scenario was, of course, taken up avidly by the cryonics community, many of whom had been closet downloaders ever since reading about intelligent robots in science fiction. Back then it was only fantasy, although computer scientists—people like Ralph Merkle, for example—would tell you that it was scientifically possible, at least in principle, to simulate a human mind inside a computer. In fact he had a proof to this effect.

"If all material objects are governed by the laws of physics," Merkle's argument went, "then the brain is governed by the laws of physics. Now a sufficiently large computer can simulate anything governed by the laws of physics. Therefore, a sufficiently large computer can simulate the brain."

It was hard to see any fallacy in this reasoning. Indeed, Merkle's proof was nothing more than a crisp summation of an argument that had been given much earlier, in the 1950s, by the British computer pioneer Alan Turing. Turing had shown that if you had a big enough memory and the right program, then a given machine—specifically what later became known as a universal Turing machine—could simulate the activities of any other possible machine, no matter how complex. The human brain was only one special type of possible machine, so there was no reason why an advanced computer couldn't simulate human brains and all of their output, things like thoughts, emotions, and everything else.

Anyway, now that Moravec had come forward with a bunch of realistic, scientifically responsible ways of actually *doing* the mind-to-metal transfer, cryonicists latched on to downloading as their ultimate guarantee of life everlasting. Prior to Moravec there had always been this embarrassing snag in the whole cryonics picture. You could get yourself suspended, resurrected, and even returned to health and youth, but then twenty minutes later you could drop over dead.

A car could run into you; a tree could fall on your noggin; you could be struck by lightning. And as if that weren't enough, you could always be *murdered,* something that might be an especial danger in a society that already had so many resurrected people to cope with that they didn't want to see yet another ugly defrostee being lifted out of the cryonics tank.

But now with downloading, all those residual worries would be *over:* backup copies would always be there to save your skin.

Bob Ettinger, it's true, saw some problems with the notion that all those separate backup copies would still be *you* in some significant sense. "Duplicates, whether high fidelity or not, are distinct individuals," he said in a review of Moravec's book. "Even if they feel in identical ways, each brain has a separate existence, and can die a separate death."

For that matter, Ettinger was not as confident as Moravec was that the self could in fact be reduced to software.

"A particular self consists of special activity in a particular mass of matter," Ettinger wrote. "It is not an abstraction. Feeling is a physical activity. We could no doubt in principle write a description of feeling, encoded perhaps in a pile of notebooks, or in a computer store—but those notebooks and notes will not feel anything, and perhaps the computer can't either."

By and large, though, cryonicists saw such skeptical doubts as retrograde reasoning. The point was, Moravec had latched on to something that just might work. Who could look that in the eye and turn down the chance it offered, especially since you might be able to download yourself before you ever needed to be frozen? For Moravec had said, again and again, that all this could happen in a mere *fifty years.* Suddenly, the younger cryonicists saw that they might not have to be frozen at all. They could put themselves into computers while they were still alive, have backup copies made, and live on forever afterward.

Of course there were dangers even with this. You might get so seduced by simulated reality that you forgot all about the "real world." Instead of letting yourself be hooked up to external sensors, or placed into a robotic body of some type, you got addicted to synthetic experiences just as, in earlier times, people got addicted to drugs.

Arthur Clarke, indeed, had seen this possibility too, back when he was considering what those "recent developments in information theory" might lead to in the end. One thing they might lead to was a society where people lived their whole lives inside of simulated realities, wired up to "thought projectors," machines that produced mental experiences so real that "the brain would think it was experiencing reality. There would be no way in which it could detect the deception."

Later, Fred Pohl wrote about "the joy machine," a contrivance offering "a subjectively real mechanical reproduction of *any* sensation you wish." Such a machine, he stated, "might be built a century or so from now."

On the other hand, maybe people wouldn't let themselves get addicted to fake realities after all. Robert Nozick—the philosopher who claimed that humanity had *lost its claim to continue*—once wondered whether life would be worth living if you were confined to a simulated reality exclusively.

"Super-duper neuropsychologists could stimulate your brain so that you would think and feel you were writing a great novel, or making a friend, or reading an interesting book. All the time you'd be floating in a tank, with electrodes attached to your brain. Should you plug into this machine for life, preprogramming your life's experiences?"

Nozick, for one, didn't think so. "Plugging into the machine," he claimed, "is a kind of suicide. We want to *do* certain things, and not just have the experience of doing them. We want to *be* a certain way, to be a certain sort of person. Someone floating in the tank is an indeterminate blob."

The last thing a cryonicist had in mind for himself was to end up as *an indeterminate blob,* but so far as Moravec was concerned, there was not much likelihood of that happening. The alternative scenario would be far too attractive: the chance of putting yourself into a fabulous bush robot body, thereby becoming the all-powerful superman you'd always wanted to be. Indeed, omnipotence would be an everyday thing, in the Age of Postbiological Man.

*T*he Age of Postbiological Man would reveal the human condition for what it actually is, which is to say, *a condition to be gotten out of*. Friedrich Nietzsche, the philosopher, had already seen the truth of this back in the nineteenth century: "Man is something that should be overcome," he'd written in 1883. "What have you done to overcome him?" Back then, of course, the question was only rhetorical, but now, in *fin-de-siècle* twentieth century, we had all the necessary means in front of us (or soon would have if we listened to Hans Moravec) for turning ourselves into the most advanced transhumans imaginable.

As for the leftover human beings that might still be around after the great transformation, well . . . they'd be allowed to remain human, of course. They might even be valued for historical purposes, as repositories of that outmoded item, DNA. But on the other hand they might very well be relegated to zoos, museums, nature preserves, or the like.

"It would cost the robots very little to maintain a nature preserve where human beings could live," Moravec once speculated. "But you've got to remember that the human being is this very limited thing that will not be able to participate in all of the good stuff that's going to be going on in those times.

"It's like having, maybe, a dog in the city. The dog can get along in the city if you're careful, if you keep it on a leash, but it's not going to be able to work automated teller machines or drive cars or fly airplanes. And it's going to be the same thing with people. We're going to be so enormously out of our league, if we stay in these bodies and these brains, that all we can hope for is to have specially prepared areas set aside for us, like nature preserves, where we can't get into too much trouble."

These bodies and these brains . . . It was *a real pain in the neck*, this business of being confined to *these bodies and these brains*.

"I resent the fact that I have these very insistent drives which take an enormous amount of effort to satisfy and are never completely appeased," Moravec said, speaking of things like food and sex, which he enjoyed just like anyone else. But he also resented the fact that he enjoyed them. Indeed, he'd never been too keen about the human, all-too-human body that he himself had come equipped with, and his distaste for it reached a peak the time he had to have it repaired in the hospital.

It was a double embarrassment. Not only was his body failing him, proving that he too was subject to all the ills of the flesh, but it was on this occasion that he finally learned—absolutely, and *for sure*—that he really wasn't a robot.

There he was on the operating table, being cut open for renovations, and what should the doctors find but . . . ordinary flesh and blood. A small disappointment, not entirely unexpected, but still, he'd never had to give up hope till then.

On the other hand there was a silver lining here, for it was also during his illness that Moravec met his future wife, Ella. She was one of his nurses. How could it not be true love when she sees him there on the hospital bed, in full and flagrant vulnerability—nothing but blood and guts; he's *not even a robot,* for crying out loud!—and she accepts him anyway?

Moravec loves his wife, make no mistake! In fact, he dedicated *Mind Children* to three people: to his father, "who taught me to tinker," to his mother, "who taught me to read," and "to Ella, who made me complete."

Human though he was, Hans Moravec was as happy a person as you'd ever want to meet. And why shouldn't he be? He understood the human condition for what it was . . . and he could see a way out of it.

All we had to do was get rid of *these bodies and these brains.*

"I have faith in these computers," Hans Moravec said. "This is not some way of tricking you into being less than you are; you're going to be *more* than you are. You're going to be more intelligent, you'll be able to do much more, understand much more, go more places, not die—all those things.

"It really is sort of a Christian fantasy: this is how to become pure spirit."

6

The Artificial Life
4-H Show

Well, all that fine theoretical talk about omnipotent assemblers performing miracles of creation *(complete control over the structure of matter)*, the talk about people putting themselves in computers and then installing those computers in bush robots *(in just fifty years)*, all that talk was well and good—except that a skeptical type might be permitted a brief moment of wondering what had been accomplished so far.

What progress had been made? How much of those grandiose plans had actually been achieved? There had to be some anchor for this in the real world, otherwise all that top-notch hubristic adventuring, as admirable as it might have been in its own right, all of it could be dismissed as . . . mere *theory*.

The fact was that the required Show 'n' Tell for the hubristic intellectual voyagers actually took place. It began on Monday morning, September 21, 1987—that's when it started, the world's first conference on artificial life. How fitting it was, too, that the conference was held at the Los Alamos National Laboratory, a place that, for obvious historical reasons, was pretty much given over to *results*. You could talk all you wanted, at the lab, about what type of explosive might be best suited for this or that sort of bomb, but then you had to go out to the canyon, or to the proving ground, or over to the Nevada test site, and put your invention to actual trial. Hopefully, it would blow up.

That was what counted at Los Alamos—*results*. Of course there was a measure of irony here, which was not lost on the participants. As Chris Langton, the conference organizer, put it: "Here we are at Los Alamos, the home of the atom bomb. This is where we developed the technology of mass destruction and death, and it's also the home of the first artificial-life workshop. And just like the development of nuclear power, there's equal chance for use and abuse of artificial life."

Anyway, by Sunday night, September 20, most everyone seemed to be there. They came from just about every state in the union, although there was a disproportionate number of attendees from the high-tech states of Massachusetts, Arizona, and California. And they came from abroad: England, Denmark, the Netherlands, West Germany.

When these people looked through the program for the workshop, officially billed as "an interdisciplinary workshop on the synthesis and simulation of living systems," it became clear at once that there were going to be lots of artificial living beings doing their artificial things in the next few days: artificial flowers, insects, flocking birds, schooling fish, swarming bees, to name just a few.

Richard Dawkins, the zoologist, arrived from Oxford, bringing with him his automatic "biomorph" computer program. There was Dave Jefferson, of UCLA, with his warren of artificial rabbits and foxes. Aristid Lindenmayer, from Holland, land of the tulip bulbs, showed up with beds of artificial lily-of-the-valley, lilacs, and flowers never before seen on earth or in heaven. Pauline Hogeweg, also from the Netherlands, bought along her "bioinformatic bumblebees," and Norman Packard, from Illinois, turned up with a colony of generic computer insects. Craig Reynolds, from Symbolics, a Los Angeles computer firm, came with a small flock of "boids" (bird-oid objects), and A. K. Dewdney, the Computer Recreations columnist for *Scientific American,* was there to judge the workshop's Artificial Life 4-H Show and to award the blue ribbons and prize certificates that he'd made with his own hands.

In all, more than a hundred active researchers—including Hans Moravec, Eric Drexler, and Keith Henson—showed up for the five-day affair. They were the cream of the artificial-life crop, these people, and what with all their demonstrations, simulations, models, computer programs, videotapes, and films, what with all their talk of "recent experiments in synthetic biology," of producing "living forms from inanimate matter," of "building authentic brains," well, you might think they'd be able to perform any arbitrary feat of creation, turning out synthetic organisms from scratch right there in front of you, in a test tube, perhaps, or maybe on a makeshift operating table, in the manner of that other famed artificial-life pioneer, Victor Frankenstein.

But reality did not quite match up to expectation at the artificial-life show. Other than for the people who attended, the most advanced moving entities there were a pair of little motorized vehicles ("artificial cars").

They drove themselves toward sources of light, like moths.

*F*rom the history of the enterprise, you'd think that the place would have been crawling with lots more exotic life-forms than a couple of light-seeking toy cars, for attempts at making mechanical life-forms went back a long way, to the clockwork automatons of the eighteenth century. Chris Langton, who had coined the very term *artificial life* back when he was a graduate student at the University of Arizona, gave a welcoming address in which he told about the artificial duck that had been put together by one Jacques de Vaucanson in 1735. Made out of gilded copper, the bird not only looked like a duck and quacked like a duck, it also flapped its wings, splashed around on the water, and even ate, drank, digested, and excreted pellets of artificial food. The thing was a masterpiece of complexity, one wing alone having been made out of four hundred separate pieces of mechanism.

And that was not even the world's most advanced artificial duck! A much better one was made a hundred years later by a man named Reichsteiner. When he exhibited it in 1847 the newspaper *Das Freie Wort* gave an ecstatic account of how realistic it was.

"After a light touch on a point on the base," the newspaper said, "the duck in the most natural way in the world begins to look around him, eyeing the audience with an intelligent air. His lord and master, however, apparently interprets this differently, for he soon goes off to look for something for the bird to eat. No sooner has he filled a dish with oatmeal porridge than our famished friend plunges his beak deep into it, showing his satisfaction by some characteristic movements of his tail. The way in which he takes the porridge and swallows it greedily is extraordinarily true to life. In next to no time the basin has been half emptied, although on several occasions the bird, as if alarmed by some unfamiliar noises, has raised his head and glanced curiously around him. After this, satisfied with his frugal meal, he stands up and begins to flap his wings and to stretch himself while expressing his gratitude by several contented quacks."

That was state-of-the-art artificial duckdom, circa 1847. At mid-twentieth century, by which time the prospect of creating synthetic fowl did not seem all that implausible, there was nothing at all like the artificial duck in existence other than some weirdly lifelike simulations at the various Disney theme parks and other such places. Oh, there was the story of the artificial *fly,* as recounted by Steve Ocko of MIT's LEGO/Logo Lab, but that was hardly in the same category.

Ocko used to work for the Ideal Toy Corporation, where he'd been apprenticed to the company's master craftsman. One morning when Ocko came to work the master craftsman was standing in the center of the office demonstrating his latest, and surely his greatest, invention. It was this tiny, white winged vehicle, which was buzzing around the room, just like a fly. It was swooping and swerving and doing all manner of extremely complicated aerobatic maneuvers as the master craftsman moved the lever on this little joystick he held in his hand.

Steve Ocko thought that this was incredible. You could make *a million dollars* with this thing. A masterpiece!

Unfortunately, it turned out to be just another one of the master craftsman's practical jokes. He'd taken an ordinary housefly, glued paper wings to it, and let it go. The joystick was fake and did nothing.

The august scientists who attended the artificial-life workshop, by contrast, were thinking not only of making simulations for entertainment purposes but of *creating life itself.* Not *simulations* of life, but *real, new life-forms,* entirely original living structures, ones based on something other than the same old carbon chemistry.

As to what that something else would be, the view at the conference was that it didn't much matter. What was important was not the materials of which living things were made, but the ways in which those materials were organized, the ways in which the underlying logic of life was expressed.

In his classic textbook *Principles of Biochemistry,* A. L. Lehninger once asked a question that was basic to the whole artificial-life enterprise: "If living organisms are composed of molecules that are intrinsically inanimate, why is it that living matter differs so radically from nonliving matter, which also consists of inanimate molecules?"

The answer lay in the ways those molecules were put together, for the essence of life was not substance but complex patterns of information. Life was not a "thing," but a property of the way things were organized. If you organized things correctly, then life could be made to exist in just about any substance, whether in flesh and blood, or in blips on a computer screen, or in grains of sand. The challenge at the conference was to capture the necessary patterns in something other than the usual materials of biology and botany.

There were two basic motives behind the enterprise. One was to discover a universal biology, based on more than a single example of life, which was all we'd had on earth thus far. The fact was that earthly biologists studied not life per se, but only *earthly* life, and in fact only one specific form of it at that. No matter how different from each other earthly life forms might appear—a worm from a bird, an ant from a whale—nonetheless these and all the other examples of life on earth had descended from a single genetic source and hence were all examples of only one possible way of making living things. As Carl Sagan had once put it, "The biologist is fundamentally handicapped as compared, say, to the chemist, or physicist, or geologist, or meteorologist, who now can study aspects of his discipline beyond the earth. If there is truly only one sort of life on earth, then perspective is lacking in the most fundamental way."

Chris Langton saw the problem in much the same terms. "Without other examples," he said, "it is extremely difficult to distinguish essential properties of life—properties that must be shared by any living system *in principle*—from properties that are incidental to life, but which happen to be universal to life on earth due solely to a combination of local historical accident and common genetic descent."

Sagan's solution to the Only-One-Example-of-Life problem was to search the skies for alternative life-forms, which is to say, extraterrestrials.

"The possession of even a single example of extraterrestrial life," he said, "no matter how seemingly elementary in form or substance, would represent a fundamental revolution in biology."

Chris Langton, though, saw little hope there. "Since it is quite unlikely that organisms based on different physical chemistries will present themselves for us to study in the foreseeable future, our only alternative is to try to synthesize alternative life-forms ourselves—*Artificial Life*: life made by man rather than by nature."

So that was one motive behind the artifical-life effort, the attempt to formulate a universal biology. But that was not the *real* reason for trying to create artificial life. The *real* reason was that creating life would be a fun thing to do, in and of itself, just as, forty years earlier, and also at Los Alamos, making The Bomb was.

"If you have a really neat idea, I think you should do it," Hans Moravec said after the conference, "because it will open new vistas for you. If you have a few pounds of plutonium—hey, go ahead! The Manhattan Project was caught up in that."

And so when the structure of DNA was discovered in 1953 and the possibility of actually creating life arose, well, what better scientific project could possibly be imagined than "life made by man rather than by nature"? Indeed in 1965, a dozen years after Watson and Crick, Charles Price (who was then the president of the American Chemical Society) suggested that creating synthetic life-forms be made a new national goal for America. These were to be *actual new life-forms*, he said, *not* "mere imitations."

Naturally, there were some risks involved, not the least of them being how to figure out what success at the enterprise might *mean*. After all, there was this whole Frankenstein business to worry about, the image of the Mad Scientist going down into the basement, mucking around in the rag-and-bone shop, and coming up many years later with a Monster.

At the Los Alamos workshop Doyne Farmer, a physicist, called this "the Frankenstein question," and said it was "the bugaboo metaphor for artificial life." But few of those present spent much time puzzling over the issue. There were all these *new life-forms* to see! What it all meant could be figured out later. Maybe.

*T*heory was one thing but actual practice was something else again. In general, there had been lots more theories of how to make artificial life than there were even close approaches to actually doing so. In 1956, for example, an article appeared in *Scientific American* called "Artificial Living Plants," written by one Edward F. Moore. The author proposed building a series of man-made plants, machinelike entities that would use nature's own raw materials, plus sunlight, to make copies of themselves. They'd be "living" plants, he said, because they'd actually *do* what plants did, reproducing themselves after their own fashion. But they'd also be "artificial" plants inasmuch as they were man-made.

You'd make such a plant, the author said, for its economic value. "It could be harvested for a material it extracted or synthesized, just as cotton, mahogany, and sugarcane are now harvested from plants in nature. Thus an artificial plant which used magnesium as its chief structural material could be harvested for its magnesium."

Except for being on the macro instead of the micro scale, Moore's scheme was much like Eric Drexler's. Moore envisioned making the plants out of actual machine parts—"ferromagnetic materials, electric motors, machine tools, gears, screws, wires, valves, and lubricating oil"—and he anticipated, as Drexler did, that if left to their own devices these things might become dangerous, that they'd "soon fill up the oceans and the continents," a variety of Drexler's gray-goo problem. But despite this risk, and the difficulty of actually creating artificial living plants, Moore speculated that "the whole design problem could probably be solved in five or ten years." However long it took, "the achievement would be more easily attainable than human flight to other planets in a spaceship."

That was *theory.* At the Los Alamos workshop, Richard Laing, a computer theorist, envisioned putting fleets of such self-reproducing machines on the moon and turning them into automated factories, giving us all sorts of manufactured products for "almost nothing."

That too was theory. As for actually *making* any artificial life, progress had been considerably more limited, although even here there had already been some small successes. In 1969, the year of the moon landing, a group of three biologists, K. W. Jeon, I. J. Lorch, and J. F. Danielli, of the State University of New York at Buffalo, created something *like* an artificial living cell. Or at least it was *partly* artificial.

The three had gotten their inspiration from Charles Price's 1965 proposal that creating new life-forms be made a national goal. "After participating in a symposium on the experimental synthesis of living cells," they wrote in their report on the project, "we decided that we had the means to carry out the reassembly of *Amoeba proteus* from its major components: namely, nucleus, cytoplasm, and cell membrane."

So they took some living amoebae, tore them apart limb from limb, and out of these disattached appendages fashioned a completely new living entity. They took a membrane from one cell, cytoplasm from another, and the nucleus from a third, and then they put these things together to create a new animal.

Lo and behold, this new combination of elements *lived.* Not always, but 80 percent of the time, the new cell survived.

"We now have the technical ability," they concluded, "to assemble amoebae which contain any desired combination of components."

Jeon, Lorch, and Danielli were not actually synthesizing new life but only, as they described it, "reassembling living cells from dissociated components." But as small an achievement as that may have been, these people had created something altogether new under the sun. It was as if, on a larger scale, they had put together a human being by combining the blood system of one person with the musculature and skeleton of a second and the brain and internal organs of a third. What had they created, after all, but . . . *Frankencell!*

As to the point of it all, the newly assembled amoeba, they said, would be "an excellent test system" for the viability of other cell parts. So if for some reason you wanted to determine the health of a given cell nucleus, you could transplant it into one of their newly assembled test-bed cells and see if it lived or died.

By the time the Los Alamos workshop got going in September of 1987, all this stuff was far in the past. Nevertheless, so far as actually producing any new life-forms was concerned, there had not been a whole lot of progress in the interim. Not so much as a single synthetic cell had been made from scratch, not one artificial living plant had sent out any mechanical roots. There was nothing at the workshop that even approached, much less topped, the old mechanical ducks of Vaucanson and Reichsteiner, and in fact the mismatch between theory on the one hand and actual achievement on the other struck, to some observers, a rather jarring note.

Hans Moravec showed up with his Postbiological Man scenario, telling his fellow Mad Scientists all about how his superintelligent robots of the future—"artificial people"—would start replacing the current version of humanity in *just fifty years!*

"It will be then that our DNA will be out of a job, having passed the torch, and lost the race, to a new kind of competition. The genetic information carrier, in the new scheme of things, will no longer be cells but knowledge, passed from mind to artificial mind."

Moravec told them about the triumph of downloaded computer minds blazing off to other galaxies on beams of controlled energy, and so on. All of that was pure theory and exciting enough in its own wild way, but then Moravec showed the videotape he'd brought with him. The tape was a measure of actual progress to date, and it was far from exciting.

The videotape went back to his old Stanford days when Moravec was living up in the ceiling at SAIL, the Stanford Artificial Intelligence Laboratory (there were whole student-built apartment complexes up in the ceiling), and working on his first major "adult" robot, Cart. Cart was supposed to be an "automatic car," a vehicle that could travel around on its own recognizance without destroying itself by crashing into obstacles, running off cliffs, and so forth. It didn't *look* like a car, however, as it was essentially four bicycle wheels topped by a deck bearing a TV camera for eyes and a radio antenna for sending signals back and forth to an external monitoring screen. The whole thing was controlled by a separate computer that would take the televised images, construct a three-dimensional map of the robot's current environment, and then plot out a safe course through it.

Cart did what it was supposed to do, but the problem was that it took forever. Just navigating from one side of the room to the other was a major achievement, something that in its *best* run took it a full five hours.

All of this was on Moravec's videotape—speeded up, of course, so that it only took a few minutes of actual viewing time. Nevertheless, it seemed to take the robot an eternity *just to cross the room!*

Moravec's newest robot, Neptune by name, did not do much better. It was a product of his own Mobile Robot Laboratory up at Carnegie-Mellon and was more advanced in every way. It even *looked* a little more futuristic, but Neptune too scarcely did anything more than traverse open space, albeit at a greater rate of speed than its predecessor, the slow-mo Cart.

His videotapes were on the verge of being *depressing,* but Moravec himself was not perturbed. For one thing, he still had his *fifty years* ahead of him. For another, he'd been saying all along that the actual development of intelligent robotic life would pretty much have to recapitulate the evolution of animal intelligence on earth. In other words you'd have to go from cells to lowly organisms like worms and insects before you could try to build anything as complex as an artificial person. That his robots could just barely stagger across level ground merely accentuated the fact that artificial life was still at the insect stage. (Or maybe even a little *earlier*—the time at which protoplasm was first oozing its way through the primordial slime.)

This was a fact that, by the time of the artificial-life conference, many researchers were acknowledging quite frankly. Up at MIT, for example, there was a so-called Artificial Insect Laboratory. Rod Brooks, one of the computer scientists up there, had dreamed this up. Brooks had worked with Moravec back at Stanford—he'd even made some of the Cart videotapes—and had then tried to build some robots of his own, ones that were *intentionally* designed to be as primitive as mere insects. They'd operate more or less perceptually, on the basis of immediate information fed in by their sonar or infrared sensors, much as an insect does out in the wild.

The advantage here, supposedly, was twofold: true-to-lifeness and speed of operation. Rather than build up complex computer simulations of their environment as Moravec's Cart did, the artificial insects would work in more of a stimulus-and-response mode. Indeed, when Brooks got them working (itself a rare event), the artificial insects ran around the room almost as fast as real bugs.

The question always remained, though, whether any of these robots—from Moravec's Cart to Brooks's synthetic insects—were "alive" in any reasonable sense of the term, even *artificially* alive. Rod Brooks, for one, wasn't sure. "It's not an easy question," he said. "I don't think of these things as 'robots,' I think of them as 'creatures.' When they're switched on, they *live* in some sense. It'll be easier for others to accept this when one of the creatures stays switched on for months at a time, but we're not quite at that point yet."

On the other hand there *had* been some surprises with the artificial insects. There was the night Rod Brooks was alone in his lab, the Artificial Insect Laboratory, working late, fiddling with some innards at the heart of one of his bugs. Everything was quiet, there wasn't a soul around, nothing was happening, when suddenly one of his mechanical bugs *started running!* All by itself this motorized insect was spontaneously . . . *coming to life!*

It turned out that Brooks had forgotten to shut the thing off and the robot had merely responded to one of his own random movements. Nevertheless, the episode sent shivers up his spine.

*I*f Moravec's videotaped robots were not obvious examples of living entities, some of the other alleged artificial life-forms at the Los Alamos conference were even less promising. There were the "memes," for example, as described by Keith Henson.

Henson had known Chris Langton, the conference organizer, back in Tucson, where they both went to college. Langton had been an L5 Society member and like everyone else used to come over to the Hensons to "help out," but then gradually all of them went their separate ways. Back then, according to Henson, they had both been infected by the space colony "meme"; now they'd been infected by the artificial-life "meme." There was nothing mysterious about this, memes working, in Keith's view, much as real-world viruses did.

The meme concept itself went back to Richard Dawkins's book, *The Selfish Gene*. Dawkins had noticed that genes were not the only things that evolved according to the principles of natural selection; so did patterns of information of every other type. *Ideas,* for example, evolved in quite literal senses: they replicated when they were communicated from person to person, they mutated when different people made their own alterations, and finally—when they'd outlived their usefulness—they became extinct. Beliefs, social practices, fads, all these could be considered to be "memes," a term Dawkins coined as an analogue to "genes."

Henson told of the time he himself was personally responsible for the *birth* of a meme, which then underwent diffusion, mutation, and ultimate extinction, all in classic biological fashion. This was when he first registered at the University of Arizona, back in Tucson. There was a punch card for religion in his packet of registration materials, something that profoundly offended Keith Henson. "I figured that they would sort this card out and send it to the 'church of your choice' so the church could send around press gangs on Sunday mornings."

But Henson was not particularly religious, so he put down "MYOB" on the card, for Mind Your Own Business. The next semester, though, he had a better idea. He put down "Druid." There were registration checkers who looked over your forms before you handed them in, and when he was asked to explain "Druid," Henson launched into a long, canned diatribe about how the Druids had been around for much longer than the Christians, et cetera, and the checker rolled his eyes and waved him on without further argument.

Of course this was too good a gimmick to keep to himself, and in short order word got around the campus about how to fill out the university's religious-preference form. Henson's "Druid-registration-behavior meme" went from student to student until, at one point, a full 20 percent of them were listing themselves as Druids.

There had been mutations, however, and so there were *Reformed* Druids, *Zen* Druids ("They worshiped trees that may or may not have been there," Henson said), *Latter-Day* Druids, and so on.

"This memetic infection was faithfully passed down from year to year infecting the incoming students," Henson said, "many of whom thumbed their noses in this small way at the administration for the rest of their college years."

The Druid meme became extinct when the college administrators finally removed the religion question from registration forms. The whole experience illustrated how ideas can and do behave much in the manner of genes, replicating, evolving, interacting with the larger environment.

None of this was so surprising in retrospect, in the wake of Claude Shannon, Alan Turing, and Watson and Crick. The genes were information carriers, bearers of programs. Richard Dawkins had seen this quite vividly one day while he was watching seeds fall from a willow tree. What was the tree doing, he realized, but raining *programs*. "It is raining instructions out there," he thought. "It's raining programs; it's raining tree-growing, fluff-spreading algorithms. That's not a metaphor, it's the plain truth. It couldn't be any plainer if it were raining floppy disks."

Regarding whether memes were enough like genes that they could be considered *actual living things,* Keith Henson apparently thought so. "These memes that make up our culture are essentially living entities," he said. "They struggle against each other for space in minds and lives, they are continually evolving. New memes arise in human mental modules, old memes mutate."

Others had been even more emphatic that memes were as alive as anything else. Back in the mid-1970s, when Dawkins was just coming up with the concept, one of his colleagues, N. K. Humphrey, read a draft of Dawkins's "meme" chapter.

"Memes should be regarded as living structures, not just metaphorically but technically," Humphrey said afterward. "When you plant a fertile idea in my mind you literally parasitize my brain, turning it into a vehicle for the meme's propagation in just the way that a virus may parasitize the genetic mechanism of a host cell. And this isn't just a way of talking—the meme for, say, 'belief in life after death' is actually realized physically, millions of times over, as a structure in the nervous systems of individual men the world over."

As true as all this was, memes were not exactly what most of the Los Alamos people had in mind when they talked about *creating artificial life*. Memes, after all, were not discrete individual objects: they were nothing you could hold in your hand. They were replicating information patterns, true enough, but these things had been around since the dawn of human thought, so they could scarcely be considered novel man-made life-forms.

But if memes were weak examples of artificial life, things got even worse when Eric Drexler got up to give his nanotechnology talk. In nanotechnology, if anywhere, you'd have thought that artificial life would spring up and blossom. Drexler's whole scheme, after all, came out of molecular biology: his entire plan lay in the idea of replicators, tiny molecular structures reproducing themselves after the manner of biological cells. Nonetheless, Drexler didn't think his little engines of creation would really be living things: they'd only be *machines*.

And as it turned out, he had plenty of reason for thinking so. For all the similarities between biological cells and Drexler's molecular replicators, there were just too many differences between the two for the assemblers to be classified as living entities, or even *artificially* living entities. Consider, he said, what the assemblers will be made of, namely, quite conventional machine parts such as gears, bearings, electric motors, drive shafts, and so forth. Nothing very biological there. And whereas biological cells can adapt to different conditions by stretching, bending, and reshaping themselves, nanoreplicators wouldn't be able to do much of that because of the fact that their parts were rigid geometrical objects that could work only if they retained their original shape, structure, and function.

Matters were just as bad, he said, when you thought of the ways in which the two kinds of entities got their energy, raw materials, and programs. Biological cells made use of what Drexler called "diffusive transport," a system in which materials floated around at random and were picked out by the cell when needed. But the nanoreplicator wouldn't operate that way; rather, it would use "channeled transport," with materials and energy brought to it by conveyor belts and pipes, wires and cables. Nothing very biological there, either.

And finally there was the fact that nanoreplicators wouldn't be able to evolve, whereas biological cells obviously can and did. The reason for this lay in the differing structures of the two. In order for evolution to take place, the object in question had to change itself in an integrated and structurally consistent fashion. You couldn't make just *one* change, because of the fact that a living thing was an organized system and an alteration in one part of the system had to be paralleled by a corresponding alteration elsewhere in the organism. Cells could make such changes because their parts were adaptive and flexible: they could stretch, bend, and so forth, but the same was not true of the nanoreplicator's rigid framework. For a nanoreplicator, if one part was out of place, the whole thing would wind up not functioning, a circumstance that made it almost impossible for it to evolve.

The upshot of all this so far as Eric Drexler was concerned was that his little nanotechnological marvels would *not* in fact be living entities.

"Nanoreplicators will differ from organisms in such fundamental ways that it would be misleading to describe them as living things," he said. "It is entirely accurate to call them machines."

The whole thing was a bit incongruous: here comes this forward-looking thinker to an artificial-life conference and what does he tell the waiting masses but that his advanced, self-reproducing entities are most certainly *not* alive.

Die-hard machine fanatics in the audience were not dismayed, however. Hans Moravec, who of course had always *preferred* machines to biological organisms, thought that Drexler was far too conservative about what his nanoreplicators were and what they could accomplish if given a free rein. Moravec had said as much when he reviewed Drexler's *Engines of Creation* for *Technology Review,* back in 1986.

Nanoreplicators were *already* the same as cells, he said, because "living organisms are clearly machines when viewed at the molecular level." The further question of whether the nanoreplicators were in some sense "alive" was not, for Moravec, as important as the question of what the nanomachines would in the end *become*. Moravec saw them as potentially far more intelligent than mere people, and disliked Drexler's proposals for keeping them firmly under human control.

"Why should machines millions of times more intelligent, fecund, and industrious than ourselves exist only to support our ponderous, antique bodies and dim-witted minds in luxury?" Moravec wrote. "Drexler does not hint at the potential lost by keeping our creations so totally enslaved."

As for the gray-goo threat, Moravec was little more than just amused. "Drexler proposes that humanity develop a standing army of tame nanomachines that would function as a hypersophisticated immune system to defend against such outbreaks of nano wildlife. I have an image of molecular gestapo agents checking identity papers and summarily executing suspicious characters."

There was no meeting of the minds at the Los Alamos Artificial Life Workshop on the crucial issue of whether Drexler's nanomachines, or indeed any of the other ostensibly living systems discussed or displayed there, were in fact alive. There wasn't even much agreement as to how to *tell* when one of them would be alive or not. Gerald F. Joyce, from the Salk Institute in San Diego, probably had the best idea.

"If your organism comes out and says it's alive," he said, "then you're on the right track."

So far, all was theory. Nanomachines, downloaded postbiological minds, artificial living plants—not one person had ever laid eyes on any of these hypothetical entities, but matters were different when it came to the simulations presented at Los Alamos. At least you could *see* these things, right there on the computer screen.

There was no shortage of computer simulations at the artificial-life conference, but on the other hand simulations had the drawback of being "alive," if at all, only in a derivative sense of the term: they'd be *models* of living systems rather than living things themselves. But then Chris Langton, the conference organizer, had tried to blur even this distinction: "We would like to build models that are so lifelike that they would cease to be models of life and become examples of life themselves."

Langton, though, had a way of seeing artificial life everywhere, ever since he'd come up with the term *artificial life* as a graduate student. He'd grown up near Boston, where his father, a physicist, worked for a precision instrument company. Langton worked for a while as a computer programmer and had access to an early copy of John Conway's Game of Life, a program in which various patterns on the computer screen gave rise to even more patterns, some of which proliferated endlessly, others of which died out at once. This was Langton's first brush with what he now calls "propagating information structures." He got more intimately involved with such structures as a result of crashing his hang glider.

He'd been on his way south to attend the University of Arizona, driving with a bunch of hang-gliding friends, and they'd decided to fly off of every decent-looking hill they came to. They arrived at Grandfather Mountain in North Carolina, a privately owned, 5,964-foot peak, and got the owner's permisson to fly off it, which they did, several times. There was a snack bar and gift shop at the top of the mountain, and their flights were attracting so much attention that the owner of the place asked Langton and friends to keep on doing it. It was good for business, he said, and in fact he ended up offering them twenty-five dollars a day for every day they flew.

Well, what hang-glider pilot in his right mind could resist? It was a great mountain to fly from, especially when you got paid for it, and soon Langton was making a whole bunch of flights every day of the week. He did this until the day before he had to leave to register at Arizona. That was the day he crashed. He hit the ground in a squatting position, his legs coming up to his face and forcing his knees into his eye sockets.

"Mashing into the ground shook a whole bunch of neurons loose," Langton recalled long afterward. "I was in and out of consciousness while I was on the ground, and it was interesting to feel my consciousness sort of bootstrapping itself up, going out, then coming back up again."

He was in the emergency room for eight hours, where they found that he'd broken thirty-five bones, including both legs, his jaw, and the orbits of both eyes. He spent the next five months in the hospital, missing out on his first semester at Arizona.

It was while he was in the hospital—drugged with what seemed to be like fourteen different things at once—that he'd had *The Visions*.

These were not your *normal* visions, none of the usual fare such as pink elephants, ghosts, or images from your past life. No. What Chris Langton saw was something else; he saw these . . . *propagating information structures!* They were proliferating through neural space, traveling down his multiple synaptic pathways, and exploding into his mind like fireworks.

"They kept going through in my mind at times in a feverish kind of way. Without any conscious attempt to think about it, my mind was taking off and exploring all kinds of stuff in very unscientific modes. It was all on the edge of drug-induced, weird fantasy. I saw these bizarre visual pictures . . . I had rational insights . . . I didn't know what to think about it."

Later, after he'd recuperated, he finally made it to the University of Arizona. There he chose a double major in philosophy and anthropology, all the time keeping his eyes open for relevant connections to other subjects. But now, after The Visions, he'd gotten some new perspectives on things. His mind seemed to be broader, more open. He saw hidden meanings and significant relationships all over the place.

Most college teachers know the type quite well: they wander into your office and start talking about Aristotle and Darwin and Einstein and the cultivation of maize and galactic structure and transformational grammar and the Spanish Civil War and Freud and Kropotkin and the fertility rites of the Ik in Uganda, and all these things are somehow supposed to come together in the student's own original, all-embracing theory, a totally encompassing explanational schema that no one else in the world could possibly comprehend or even wants to hear about. One might offhand think that the purpose behind all this is to show off in front of the professor. It isn't. The purpose is simply to explain the world in a way that, to the student, takes due account of all the subtle patterns and fleeting concordances that he or she has observed throughout all these otherwise wholly distinct phenomena. Often enough, such students are crackpots, as is revealed by a peculiar telltale glint in the eye. Very rarely, one of them knows exactly what he's talking about.

Chris Langton was one of these latter, although one has to admit that few of his professors at Arizona saw it that way at the time. After he was well along in his graduate work, he proposed what he thought would be an entirely reasonable Ph.D. project, one that brought together certain theories in computer science, anthropology, and philosophy. He wanted to combine insights from all those disciplines, suggest an analogy between DNA and natural language as the medium by which information is transmitted, and then offer his own original theory explaining the commonalities between evolution in animals, human belief structures, and culture. He even proposed an umbrella term for the common element that he discerned among all these various phenomena: *artificial life.*

The faculty was not quite ready for all this, and Langton ultimately left Arizona for the University of Michigan, where he changed his subject and wrote a Ph.D. thesis called "Computation at the Edge of Chaos." He won a fellowship at the Center for Nonlinear Studies at the Los Alamos lab, and it was while he was there that he came up with the idea for the world's first artificial-life conference.

By this time he had seen even more of those subtle, fleeting concordances. There were all these new scientific fields out there—self-organizing systems, cellular automata, neural networks, studies of emergent behavior, and so on—and it had to be more than mere coincidence that what all of these new-wave phenomena had in common were the basic properties of living entities: the ability to generate order amid chaos, to self-organize, to reproduce. It was as if the underlying logic of life was somehow asserting itself in all these different manifestations.

Clearly, a major conference was in order, and who better to put it together than Chris Langton himself, the man who first saw the number of the beast? And thus it was that all those artificial plants and animals showed up at the Los Alamos conference. About a year or so afterward, the conference proceedings were published and a whole new scientific discipline had been born, Artificial Life.

Langton's own entry in the workshop, admittedly, was not an

explicit attempt at creating an artificial animal. It *was* a species of living thing, however, although more like a virus than anything else. Langton's creation was the world's simplest self-reproducing structure, a figure that reproduced itself on the computer screen. Looking much like the capital letter *Q*, the descender at the bottom of the letter would snake out and curve around into another *Q*, whose descender would then grow into still another letter *Q*, and so forth, until the whole screen was covered with a family of proliferating *Q*s.

Well, that was a technical achievement, certainly, a clever bit of programming, as it took some eight hundred rules for Langton to get this behavior going. There was even some rough analogy there to the way in which animals and plants reproduced themselves, by following out the complex rules stored in their genes. But as to the question of whether these proliferating *Q*s were in fact living structures of any sort, of this there was ample room for doubt.

The crux of the matter was the philosophical question of what *any* kind of life, in essence, really was. In the late 1960s Carl Sagan, who had himself done research in the life sciences, wrote the article "Life" for the *Encyclopaedia Britannica*. He surveyed the different conceptions of life studied by biologists, biochemists, ecologists, ethologists, embryologists, and so on, and concluded that "despite the enormous fund of information that each of these biological specialties has provided, it is a remarkable fact that no general agreement exists on what it is that is being studied. There is no generally accepted definition of life."

To Sagan, even the ability to self-reproduce was not a defining characteristic of life since many hybrid species, such as mules, weren't able to reproduce themselves, while some nonliving substances, such as crystals, were. Nevertheless, the ability to replicate themselves was such a common attribute of living things—a mule's individual *cells* replicated—that even quite simple self-reproducing structures (such as Langton's loops) would have as good a claim as any to be counted as "alive" in some abstract, formal sense of the word.

Then, too, the fact that Langton's loops existed only inside a computer didn't mean they weren't really alive in their own special way. There was the whole question of computer simulation to be considered: If you could simulate, on a computer, the very same behavior that went on in the external world, wasn't the simulation somehow just as good as the original?

About a year after the conference this question was considered in detail by Frank Tipler, the physicist who had coauthored *The Anthropic Cosmological Principle* with John Barrow. Since you could simulate anything you wanted, he said, suppose you simulated an entire city. "In principle," he said, "we can imagine a simulation being so good that every single *atom* in each person and each object in the city and the properties of each atom having an analogue in the simulation."

So you have this absolutely perfect simulation of a city and its inhabitants. "The key question is this," Tipler said. "Do the simulated people exist? As far as the simulated people can tell, they do. There is simply no way for the simulated people to tell that they are 'really' inside the computer, that they are merely simulated and not real."

If that were true then not only Langton's loops but all other computer simulations of living things would have to be regarded as analogously alive as their real-world counterparts. The artificial flowers at the conference (which won first prize at the Artificial Life 4-H Show)—were at least *artificially* alive.

These flowers, ferns, and other plants had been created by Aristid Lindenmayer and Przemyslaw Prusenkiewicz. It was amazing the way they quite literally *grew* in front of you on the computer screen. The simulated plants were not drawn, but had been *grown* in stages, by the computer's following out lists of mathematical instructions that imitated the way chemicals inside living plants controlled the processes of branching, budding, leafing, and flowering.

Higher up in the artificial-life kingdom were Craig Reynolds's "boids." Reynolds was a computer graphics and animation specialist at Symbolics, Inc., of Los Angeles, where he was working on a computer-animation film that was supposed to have lots of birds flying around in the background. He wondered if there was any way to create herds of animals—schools of fish, flocks of birds, and so on—without tediously calculating the trajectory of each individual member at every step of the way. So he thought to himself, why not create some computer birds, let them loose, and see if they'd flock together all by themselves?

It was a crazy idea, but on the other hand Reynolds knew that the synchronized flying of real birds was not produced by any master controller or Head Bird giving commands to all the rest. Rather, each member of the flock flew solely on the basis of its own individual perceptions of the world and of all the other birds in the assemblage. So he wrote a program that would produce "bird-oid" objects (or *boids*), and gave them a few general instructions about how to fly: Avoid colliding with other boids; match heading and speed with the others; stay together in a group.

Reynolds didn't know precisely what to expect when he let his boids fly off in computer space. Maybe they'd arrange themselves in a grid as if they were wired together. Maybe they'd line themselves up single file and fly through the air in a thin stream. What *actually* happened was that the boids gathered themselves into a flock and flew precisely as flocks of real birds did, expanding and contracting in quite natural ways.

There were a couple of unexpected bonuses as well, for not only was *the flock* behaving in a most birdlike manner, so were some of the *individual boids* themselves. At one point, a solitary boid left the pack and flew some distance away. It curved around in a loop and then—as if it realized its error—raced back to rejoin the group. There was nothing in the program that explicitly called for this, and for all intents and purposes the boid seemed to be acting on its own initiative.

And then another of the boids, *contrary* to its program, *did* smack into an obstacle. What happened at this juncture was a little unnerving: the boid bounced back from the obstacle, fluttered for a moment as if it were stunned, and then zoomed off to catch up with the flock. Reynolds never programmed *that* behavior into his boids either, but there it was.

*I*t had always been a mark of the truly successful Mad Scientist that what he created destroyed him in the end—or at least got completely out of his control. In *The Island of Dr. Moreau*, for example, H. G. Wells (who was a trained biologist) depicted a scientist trying to turn animals into people. For a while it worked: Dr. Moreau had invented a race of pseudohumans. But in short order they reverted back to sheer animalism and turned against their creator, killing him. Same thing with Victor Frankenstein and his monster. Hans Moravec, of course, foretold that his superintelligent robotic mind children would end up supplanting mere humans (which Moravec saw as a good thing). One way of monitoring the progress of the artificial-life crowd, therefore, was by checking to see who had created life-forms that . . . *had gotten out of hand*.

Precisely this had happened to Dave Jefferson, of UCLA. Jefferson was a youngish computer scientist already well known in his field for his "Time Warp" algorithm, a technique for blazing through certain otherwise interminable computable functions, when he suddenly got sidetracked into biology. He'd read Richard Dawkins's book *The Selfish Gene* and was particularly struck by the chapter on memes.

Ideas were memes, Jefferson realized, but so were computer programs. Programs, after all, *replicated,* because you could make a copy of any program quite easily. And they could also *evolve:* you could take a program, build in a random-number generator so as to produce subtle mutations within the program, and let it run. Most of the time this would sabotage the program completely, but sometimes the program would run a little differently, maybe even better than it had beforehand. This too was like biology, for although most mutations were *non*adaptive, some few of them *were* adaptive. Some programs, furthermore, interacted with other programs and with a shared environment, also in the manner of animal species. You could even make programs that learned from experience.

So if computer programs could do all these things—if they could replicate, mutate, interact with an environment, and learn—then why couldn't they be considered in some sense *alive?* Why, in fact, couldn't a certain, specific bunch of programs be considered *animals* of a sort?

"I don't know how the idea hit me," Jefferson recalled later, "but one day I realized that it was possible to build a system in which a population of animals was represented as a population of programs."

So Jefferson created this computer program—"Programinals," he called it, for programmed animals—in which there'd be a whole population of self-sustaining, self-reproducing computer creatures. The various programinals would have real characteristics—ages, weights, metabolisms—and real things would happen to them: there'd be births, deaths, even extinctions. There'd be predator and prey relationships, just like with the animals out there in nature: the predators would be foxes, the prey rabbits. There'd even be a food chain, with a separate program for growing the blades of grass that the rabbits would eat. Finally, both the foxes and the rabbits could *evolve*—the rabbits could develop faster running behaviors, for example—all of which was controlled by random-number generators that caused mutations.

Artificial animals in an artificial habitat—that's what Dave Jefferson put inside his computer. So he set all this apparatus in motion and sat back to behold events.

"One of the very first things I wanted to do was duplicate the classic predator-prey relationship. You'd expect in the simplest such situation that the population of rabbits would rise, and then slightly later, out of phase, the population of foxes would rise because they'd eat a lot of rabbits, and so forth. If they eat too many rabbits, the population of rabbits falls, and then there are so many foxes that they begin to starve to death because they don't have any food. So you'd get this classic cyclic behavior in both populations, with the populations rising and falling with the same frequency, but slightly out of phase with one another."

That's what Jefferson fully expected to see happening as he ran his program. And as it turned out, that's what happened—at least at first.

"I watched the output and sure enough, I got this rise and fall. But then I got a crash: the rabbits went extinct, and then the foxes starved to death shortly afterwards."

This was *not* supposed to happen. He must have made a mistake in the programming, a common enough experience in the computer world. So he printed out the entire code and checked it through, looking for what could have gone wrong. There was nothing amiss that he could see, but just to be on the safe side he made a few minor adjustments. He thought he could tweak the code just a bit here and there to *ensure* that the classical predator/ prey rise-and-fall pattern would appear. So he made the changes, sat back, and ran it again.

Only it didn't work. "Sometimes the foxes would go extinct, and then the population of the rabbits, having no enemies, would soar, until it reached the carrying capacity of the system. Try as I might, I couldn't adjust things so that I got a stable oscillation of the two populations. Always within one or one-and-a-half cycles I would get a crash."

What was going on here? Jefferson began to make up scenarios as to what could possibly be happening to produce the anomalous output he kept on getting. He no longer thought in terms of Lisp code now, but rather in terms of rabbits and foxes.

"Maybe the foxes only eat *baby* rabbits," he reasoned. "So I'll make it so that they can't eat mature rabbits, who after all can probably run fast enough to escape."

So he'd program that in . . . but it wouldn't work.

"Or maybe they only eat *old* rabbits."

That didn't work either.

"Or *old and young* rabbits. Or maybe the rabbits have *holes* that they can hide in. I don't know. I tried to complicate the picture enough so that I'd get a stable oscillation. But *nothing worked! Always, always,* I got one or another form of extinction."

So here was Dave Jefferson, masterful prizewinning computer programmer, professor of his science at UCLA, with his very own program . . . *getting out of hand!* Here was the creation *overwhelming the creator!*

Then he had the good luck to see a paper by theoretical biologist Robert May, "Stability and Complexity in Model Ecosystems." According to the author, the so-called classical rise-and-fall population pattern occurred only in exceptionally rare instances, where everything else in the surrounding ecosystem was stable. Otherwise it didn't occur. If there were any unstable factors present, as there almost always were, you'd get other results, including mass extinctions, in very short order.

In fact there *were* unstable factors in Dave Jefferson's program: the random-number generators.

"I was finding through my program," he now realized, "something that serious biologists had known all along."

Later, when he added other features to the program, thinking that the animals would respond in a certain way, he learned some other hard lessons about animal life in the real world. He'd make an alteration to the program, thinking, for example: *Surely this gives the animal a competitive advantage. They'll multiply at the expense of animals without that ability, and the gene for it will spread in the species.*

And then, often as not, what he expected to occur was the direct opposite of what actually happened.

"Time after time, what I expected to happen didn't happen. I was no better than *an ape* at figuring out what would happen! Even when I very carefully decided that a specific attribute was going to win, I was no better than random at predicting the outcome. Half the time I'd be wrong, and then I'd have to sit there and struggle with the question of *why*. So I'd run the simulation back and I'd instrument it differently and I'd watch it again. Usually it would take an hour to do the simulation and then it would take days—or even weeks—to figure out why my expectation was wrong."

One time, he put two identical species into the environment and made them natural enemies of each other.

"This was not a terribly realistic biological assumption. I don't know of any actual co-predator relationships; they're at least very rare. But anyway, I invented two such species, and they operated according to the following principles: if two animals of these opposite kinds fought each other, then the bigger one would win the fight, unless there were fights of three against two, or something like that, in which case the larger gang would win the fight."

Here, he thought, you could predict what would happen *for sure*. If you had two equal species fighting each other, then both ought to suffer the same fate. They *both* ought to die out, or they *both* ought to prosper, because both of them started off equal and had the same behavior patterns. It was hard to imagine any other possible outcome.

So of course being absolutely sure of what would happen, Jefferson found that the very opposite thing occurred. One species was not only systematically overtaking the other, it was quite literally running rings around it.

"There was a striking geometry to all of this: the one species would take over a little region of space and then would grow at the perimeter and then expand out in ever-widening rings. Meanwhile, in the center of the ring, the other species remained stationary. But as to why this was happening, I couldn't figure it out."

And then suddenly—*Yes!* . . . It *had* to be!—an adaptive mutation must have occurred!

So Jefferson looked back into the record of what had happened, and sure enough, there *had* been such a mutation. It was an extremely strange mutation, for now one of the species *couldn't move!* It was as if those animals were stuck onto rocks like barnacles. The other species could still move about normally, and so of course this should have explained everything. The stationary species would be eaten alive by the animals that still moved around.

But that's not what happened!

What happened was that the *immobilized* species survived, while the moving species was being exterminated!

"It was so counterintuitive!" said Jefferson, "*Not* having the ability to move turned out to be a *winning* genetic mutation! This astonished me. I had to take a week to figure out what the hell was going on, and I ended up running this program hundreds of times."

At length, though, Jefferson reasoned it all out.

"The essence of it was that whoever moves takes the risk of moving into the visual field of his enemy. If you *don't* move, you *don't* take that risk. And if there are enough of your enemies around so that they come to you at regular rates, you'll actually prosper. Your food will come to you."

It was a rather strange way of learning about nature. There was Dave Jefferson sitting in his office and learning about the birds and the bees *not* by reading zoology textbooks and *not* by doing experiments with animals in cages, but by trying to second-guess the output of his own computer program. He'd consult the zoology textbooks only to verify what his computerized animals were actually doing to each other down there in the silicon chips.

As for the issue of whether his programmed animals were actually "alive," that was a question to ponder, all right. Dave Jefferson once wrote out his answer to this.

"I would not hesitate to say that a program can be 'alive,'" he wrote. "Whatever reasonable definition one gives of 'life' (e.g., an energetically open system that adapts to its environment and produces variant copies of itself), there will be programs that satisfy the definition. Any claim that a program cannot be alive reflects either too narrow a definition of 'life,' or too impoverished a vision of the richness and variety of computation.

"There is more than one level of life; we can recognize at least three: the cell, the individual multicellular creature, and the population. A cell can die while the individual it is part of lives, or the individual might die while its cells remain alive. Likewise an individual may die while the population it is part of remains alive, or the population may die (suffer irreversible ecological disorganization) while the individuals survive.

"I would NOT claim that the Foxes, Rabbits, and Grass are alive at the *individual* level. I doubt that any five pages of Lisp code deserves to be called alive as an individual. But the population of artificial rabbits surely exhibits all of the qualities of a living population. It grows, adapts, reproduces, and evolves. I can see no reason to deny that this population of artificial rabbits is alive at that level of organization."

A year after the Los Alamos conference, Hans Moravec and Fred Pohl, the science fiction writer, struck up a correspondence. Pohl had written a review of *Mind Children* for the British publication *New Scientist,* and loved the book. He sent an advance copy of the review to Moravec, telling him in a cover letter how much he liked it. Pohl had long been thinking along mind-transplant lines himself and was thankful to see someone in the hard-science world finally taking some of these ideas seriously. Anyway, the two of them corresponded back and forth about science fiction and one thing and another, and in short order matters had progressed to the point where Moravec was telling Pohl of this great story idea he'd come up with.

"It would be possible to tell interesting tales set in a superintelligent postbiological (post-Darwinian?) world," Moravec wrote. "The ecology of that world would be much richer than ours now, and there would be niches for all kinds of entities, some at a scale and with motives understandable by us. One idea is to tell it from the point of view of a clan of parasites (computer viruses) who are roughly of human intelligence, because in their particular habitat being any more complex would disproportionately increase their risk of detection and erasure (or maybe the computational crumbs that fall 'where' they live are simply too limited to sustain a larger size). Their world is inhabited by gods of unimaginable power and motives, but that's mostly beyond them; with luck they will escape notice, and they spend most of their time, energy, and thoughts merely eking out a precarious, but interesting, living in the interstices. (Sounds just like our life now, come to think of it.)"

Pohl liked the idea so much that he volunteered to collaborate if Moravec were so inclined, and before long they had a gentleman's agreement and were talking about advances, agents, publishers, and all the rest of it.

By midsummer 1989 Moravec had fleshed out a plot. The story's hero, a computer virus, wants to assemble a physical body with which to break out of its environment, escape into the external, flesh-and-blood universe, and reproduce itself. Resources are scarce inside the host computer, but somehow the virus manages the feat and emerges as an actual physical being. It then has to construct a network in which it can plant a seed of itself, thereby starting a new cycle of birth, growth, and self-reproduction, but how can a newly emerged computer virus gain the necessary information to build such a network? Easy: it raids a library and, as Moravec wrote to Pohl, "lifts a genuine human mind from a history book."

So this computer virus lifts the mind from the history book and learns that it can control that mind completely. It can rewind it, like a film. It can perform all kinds of miracles.

"The plot itself might simply be the story of a virus reproductive cycle," Moravec wrote to Pohl, "or of the rebellion of the trapped human, with mind-stretching scenery, most of it equally surprising to the virus, the human, the reader, and the authors."

Fred Pohl loved it. Moravec certainly had a mind for fiction.

And so what if there was far more *tell* than *show* at the Los Alamos Show 'n' Tell? If you accepted Moravec's view that artificial life would recapitulate the evolution of natural organisms, then you had to concede that Mother Nature hadn't done her work overnight. It took a while for her to evolve all those adaptive mutations, all those sophisticated emergent behaviors, and the rest. You couldn't exactly *hurry* these things.

Besides, back in the Mobile Robot Laboratory at Carnegie-Mellon—the world of actual practice, the world of *results*—Moravec was making great strides. He was now working on a new robot named Uranus, which was turning out to be a truly advanced specimen.

It was incredible. It had a completely new computer-vision and navigational system. It had a self-contained energy source. And it motored around on these weird "Swedish wheels" that allowed it to turn completely around within its own radius. Uranus was so dextrous and powerful that it could even go up and down slight inclines, like the one out in the hallway for people in wheelchairs to use. There was even a space on its back where Moravec, or anyone else who wanted to, could sit while the thing was running, and ride it to and fro around the room, like a toy.

So Hans Moravec and his assistants and visitors took turns sitting on Uranus, riding it up and down the wheelchair ramp, and having great fun.

In the world of actual practice—where what counted was *results*—this was real progress.

7

Hints for the Better Operation of the Universe

The artificial-life conference reminded some of another Los Alamos gathering that had been held a few years earlier, on the subject of flying to the stars—"Interstellar Migration and the Human Experience," as it was called. The main feature on that occasion was a tall and rangy Texan by the name of Dave Criswell.

Criswell was a physicist associated with the University of California at San Diego. He worked for the California Space Institute, the state's own miniversion of NASA, and prior to that had been a senior scientist for the Lunar and Planetary Institute in Houston, just outside the Johnson Space Center. He was an all-purpose physicist and had done research on topics ranging from abstract plasma physics to the acoustic properties of lunar soil to assessing risk factors in NASA's space vehicles. During the prelaunch days of *Apollo 11,* the Neil Armstrong–Buzz Aldrin–Mike Collins flight, Criswell was given the task of analyzing possible mission failure modes on the lunar excursion module. He discovered that there was a slight risk that if two adjacent thrusters vented their different types of propellants after landing on the moon, those propellants might mix together and explode some time afterward. This had been observed during lab tests in vacuum chambers on earth, never doing much damage. "The effect might have been like the college chemistry prank of detonating powder put on the floor of the chemistry lab," Criswell said much later.

It was one of a thousand small details that, somehow, had to be incorporated in flight planning. As it turned out, no such explosions ever occurred during the Apollo flights.

Anyway, it was a glorious spring morning when Criswell spoke at the interstellar migration conference. The sun, mother of us all, had risen over the snowcapped Sangre de Cristos some fifty miles to the east and was sending shafts of light down into the finger canyons on either side of the Los Alamos lab, down to where the atomic scientists used to test their implosion lenses and other of their cranky little explosive devices. He was speaking in the J. Robert Oppenheimer Study Center, in a subdued room with carpets and sound-deadening ceilings and soft overhead lighting, the better to focus attention on the lecturer. Dave Criswell is up there now, holding forth in front of a small audience of about twenty fellow scientists plus a few assorted humanists. He's dressed in a blue suit, white shirt, and tie, and with his mustache and goatee he looks very sharp indeed, like a riverboat gambler, perhaps. He's got a microphone in his hand, for the whole event is being tape-recorded, and as he paces back and forth across the front of the room he whips the microphone cord out in front of him in the manner of a nightclub comic, as if he were playing a gig at Tahoe or the Catskills. He whips it out there smartly, so that he won't stumble across it in the course of his frequent room crossings, but also, if the truth be told, partly just for the panache of it all.

For Dave Criswell is a man with style—a model of Texas swagger, in fact, what with his twang, his cowboy boots, his whipping that mike cord out in front of him—but if anyone in the audience is annoyed by this they don't show it. These metaphysical flourishes are, after all, as nothing compared to the message that Criswell is now delivering to his fellow star-voyagers. He's telling the audience of this scheme he has for taking the sun apart.

What Criswell has in mind is dismantling it, squeezing the sun inside of a vast ring of particle accelerators until the hot, glowing solar gases give up and come streaming off in great scalding bursts.

"Star lifting," he calls it, a way of appropriating some much-needed building materials and, at the same time, a means of artificially prolonging the sun's lifetime. It would be a boon to mankind, a blessing, a windfall, a bonanza of stupendous proportions.

That was the claim, but who even in *this* audience could lend it any faith and credit?

They were a bunch of advanced thinkers, this group, and by this time they'd heard pretty much everything. They'd listened to Glen David Brin, the combination theoretical physicist and award-winning science fiction writer, describe how you could catch the space shuttle's external tanks once they were jettisoned up in orbit and remake them into usable structures: "habitats, shelters, warehouses, waste dumps, or farms." That only made good sense: lots of other thinkers had proposed similar uses for the shuttle tanks, so this was something you could wrap your mind around without injury.

They'd listened to a half-baked philosophy professor, Regis by name, analyze the moral propriety of multigenerational interstellar travel. Was it ethically okay, he wondered, to place a small human population into a space ark and send it off toward some other star system, on a trip that might take as much as four hundred years? And of course yes it was, he claimed, it was really not all that much different from the original Garden of Eden situation: "Launching one is like beginning Genesis anew," he said. People listened to this and nodded their heads, and nobody had any trouble with it.

And they'd listened to Michael Hart, the Princeton Ph.D. in astronomy who also had a law degree from the New York Law School, they listened to him talk about human beings becoming immortal, colonizing the entire Milky Way galaxy, and engaging in far-flung battles of interstellar domination.

"Outside observers looking at us might well say, 'The Milky Way galaxy is a jungle!'" Michael Hart said. "'A dog-eat-dog place where only the fittest civilizations survive.' Perhaps, on some cosmic scale, they might be right."

Nobody had any trouble with that, either. In fact, they'd seen it all before, in the movies, and so most of it was déjà vu, almost ancient history. All of which only went to show what an astrophysically case-hardened audience this was.

But when it came Dave Criswell's time to speak, well, that was a different story. Even for *this* audience, Criswell's stuff was really too much to take, entirely too mind-boggling by half. In fact, nobody had the faintest idea what to make of it all, and so, much against their will, they were forced simply to let his enormous flood of sensory data, his fully integrated sound-and-light show, roll over them like a breaking wave.

There he was, the well-educated, well-dressed plasma physicist David Russell Criswell—grandson of W. L. Russell, the only physician in Rhome, Texas, population 450, where he grew up—there he was stomping across the room in his cowboy boots, whipping that microphone cord out in front of him in that mesmerizing, up-and-down sine-wave pattern, pointing to the view-graphs on the projection screen, and reporting in his stylish Texas drawl that someday in the not-too-distant future—*only a few hundred years, in all probability!*—we'd have to start taking the sun apart.

Quite incongruous, admittedly, but that's exactly what happened at the Los Alamos National Lab one fine spring morning as the mountaintop snow melted under the sun's heat.

As far-out as it was, Dave Criswell's dismantling-the-sun scenario was not, by a long shot, the most extreme of the heroic macroengineering feats contemplated by scientists in this age of *fin-de-siècle* hubristic mania. It was not at all the ultimate in conceptual bravery, although by any standard Criswell's idea had to be regarded as uniquely thought-provoking. But coming up with such schemes was not so much a matter of sheer hubris as it was a question of seeing the universe in the right aspect, of not taking the structure of the cosmos for granted, in even the slightest degree. For in the latter quarter of the twentieth century the fact was that *you didn't have to:* never more would anyone have to take nature the way it came, as if it were just an unalterable given that you had to make do with.

With certain exceptions, anything in creation could be modified or improved upon. And maybe there weren't even that many exceptions. You might think, for example, that the basic physical constraints—the so-called laws of nature, the ultimate physical parameters such as Planck's constant, or whatever—you might think that these were set in concrete and immutable. But who could say with any assurance whether that was really so? Isaac Asimov, in his capacity as science fiction writer, had suggested in his book *Fantastic Voyage II: Destination Brain* that maybe Planck's constant could be *changed,* that maybe science could put together a machine that would *compress the very structure of space itself,* so that you could shrink people (or anything else) down to as small a size as you wanted. That was "fiction," of course, but if there was anything that physicists had learned by the late twentieth century it was to be extremely wary of saying that something was absolutely *impossible.*

Not that a given improvement could be made *right now,* necessarily, but at least the various ways of going about it would already be clear. At any rate it was increasingly easy to see ways in which Mother Nature could have done things better, universewise.

By midcentury, for example, some scientists had decided that the solar system—in its current version—was *energy-inefficient.* This embarrassing fact was realized about 1960 by Freeman Dyson, the theoretical physicist. Much of the energy in the universe, Dyson knew, resided in the stars, those celestial bodies that progressively radiated themselves away to extinction. All that starlight made for a beautiful night sky, but otherwise it was just plain gratuitous. The starlight didn't *do* anything; it performed no useful work. It just streamed off to infinity and that was the end of it. This was something that a truly intelligent society wouldn't allow to happen. Changes, after all, could be made. For example, what was to stop an advanced technological species from trying to capture that energy before it went too far? Maybe you could grab it out of the sky, bottle it up somehow, and then *use* that starlight; you could make it *do something* instead of, as Dyson put it, letting it run off, "wastefully shining all over the Galaxy."

Capturing stellar energy might be difficult, he thought, but it was by no means impossible. Suppose you enclosed the star—our very own sun, for example—inside of a sphere, a casing that surrounded the solar system in all three dimensions. The sun, the planets, their satellites, and everything else would be enclosed inside this thing as if in a vast cosmic eggshell.

That would require a goodly amount of matter, it might even mean dismantling a whole planet (Jupiter, let's say), but what were planets good for if not for their raw materials? So you'd take some planet apart—by spinning it until it flew to pieces, maybe—and then you'd refashion it into a thin enclosure around the solar system.

This shell of matter later became known among astronomers, space fans, and science fiction buffs as the "Dyson sphere," and Larry Niven, the science fiction author, wrote a novel, *Ringworld,* about living on the inside surface of a modified version. Niven, who had a physics and math background, also wrote a nonfiction essay on the relative worth of the Dyson sphere as opposed to his own improved model, the Ringworld.

One of the drawbacks of the Dyson sphere, Niven said, was that if you spun it to create gravity, then the atmosphere inside would collect at the equator, making the rest of the sphere useless. You could get around this by installing gravity generators, but what if they failed?

"Life is not necessarily pleasant in a Dyson sphere," said Niven. "We can't see the stars. It is always noon. We can't dig mines or basements. And if one of the gravity generators ever went out, the resulting disaster would make the end of the earth look trivial by comparison."

Niven acknowledged that there were ways around some of these problems. For example, there was the "Alderson Double Dyson Sphere," as conceived by Dan Alderson. This was a system of *two* Dyson spheres, one inside the other: the smaller of the two was transparent, so that the sun could shine in, and if the outer sphere was built close enough to the inner one, it would hold the atmosphere in between the two.

But Niven had some even better ideas. First of all you'd forget about building a *sphere.* That in itself was a waste of good planetary materials. Instead you could build a hollow *disk* ("like a canister of movie film") that would enclose the sun and all the planets. You could spin the disk for gravity and then the atmosphere would collect all around the perimeter.

Better yet, you wouldn't even have to *enclose* the disk. "We wouldn't even have to roof it over," Niven said. "Put walls a thousand miles high at each rim, aimed inward at the sun, and very little air will leak over the edges."

So now you'd have a ring that went completely around the sun, say at a distance of ninety-three million miles, the radius of earth's orbit. This was Niven's "Ringworld." By artful arrangement of a second, interior ring, you could even control the entrance of sunlight.

"Set up an inner ring for shadow squares—light-orbiting structures to block out part of the sunlight—and we can have day-and-night cycles in whatever period we like. And we can see the stars, unlike the inhabitants of a Dyson sphere."

Indeed, there seemed to be no end to the possibilities once you let yourself regard the solar system as *raw material,* and the greater universe as a nice place to visit. Not that Niven's Ringworld would be much suited for traveling.

"The Ringworld makes a problematical vehicle," he admitted. "What's it for? You can't land the damn thing anywhere. A traveling Ringworld is not useful as a tourist vehicle.

"A Ringworld in flight would be a bird of ill omen. It could only be fleeing some galaxy-wide disaster. Now, galaxies do explode. . . . "

For that matter, they also collide with each other. Keith Henson, who was of course well aware of this fact, once came up with a method for *moving one galaxy out of the way* of another one. This was back in the halcyon L5 Society days when he and Eric Drexler had been considering the various modifications and improvements they might make to earth, the solar system, and whatever else. Henson put forward a scheme for moving the earth back from the sun (in case it got too hot).

What you'd do is this: You'd put a large collection of solar sails ahead of the earth in its orbit. These sails would be gravitationally coupled to the earth, and as the pressure of sunlight fell on the solar sails, the sail-and-earth complex would be pushed farther out, into a bigger orbit.

That was how you'd move the earth. Then Drexler had worked out a variation on this that could be used for moving a *star*—our own sun, for example. You'd hang a hemisphere of actively controlled light sails over the star and, as before, couple them to the star's gravity. Then you'd turn the sail-star system into a fusion/photon drive, allowing you to rocket the whole apparatus through space.

Finally, Henson realized that if you did this with *every* star in a given galaxy, you'd be able to move the galaxy itself.

"While we might not ever *need* to move a galaxy," Henson said, "it's kind of nice to know we can. The Andromeda galaxy seems to be headed our way at eighty kilometers a second. At that rate, it would take five billion years to get here, but it might well take most of that time to move out of the way!"

Nor was the notion of doing things to galaxies confined only to a small circle of L5 fanatics. Freeman Dyson had considered the prospect of rearranging galaxies, too, this in view of the rather thoughtless way in which the typical garden-variety specimen was laid out. "A galaxy in the wild state," Dyson thought, "has too little matter in the form of planets, too much in the form of stars."

Not enough planets. Too many stars.

Besides that, these "untamed" galaxies had stars moving about every which way, all quite recklessly. How much better it would be if some sense of rational planning were brought to the situation; for then, as he thought, "Stars instead of moving at random would be grouped and organized."

So you'd organize the Galaxy and then you'd turn some of its excess stars into planets, so that people and other beings could live on them. There were a few different ways of managing this feat. You could, for example, induce collisions among stars, triggering artificial supernovas, the leftover debris from which could then be collected and fashioned into habitable planets.

That was only one idea—Dyson had a few other star-into-planet schemes as well—but at this relatively early stage in the game the details weren't as important as the underlying principle: that without any question *the cosmos could be remade.*

*T*his was extremely bad news to some, especially to those who preferred untouched wilderness to life in downtown Tokyo. How could anyone in their right mind regard the universe as . . . so much modeling clay? There was a certain Army Corps of Engineers mentality at work here that was not lovely to contemplate, the vision of interstellar bulldozers mowing down stars and planets as if they were mere underbrush—with Freeman Dyson waving gaily from the driver's seat.

It was quite understandable, then, that when thinkers of a more environmentalist stripe looked at the idea of space colonization from a moral or aesthetic standpoint, the whole enterprise took on a decidedly different hue. Foremost among these critics was David Thompson.

Thompson was a zoology professor and environmentalist at the University of Wisconsin Center at Washington County, a small college of about five hundred students not far from West Bend. He taught an environmental studies course on the campus, and one of the things he always tried to communicate to his students was a sense of just how fragile closed ecosystems actually were. In fact he gave them what he thought would be an unforgettable object lesson in the subject.

At the beginning of the semester Thompson would have his students collect samples of pond water, which they'd put in plastic tissue-culture flasks. He'd have the students check on the progress of these mini-ecologies at the start of every class session. They'd look through the sides of the bottles with low-power microscopes and study the various life-forms: algae, rotifers, unicellular animals.

It was hard to predict what would happen to them. Some of the students thought that the organisms would die in a matter of hours or days, but most of the time they didn't: they kept on living and reproducing for weeks. Finally, though, after about two months the environments crashed and many of the organisms died after all. Mostly what was left at the end were rotifers, and they cleaned up the dead bodies of the other animals.

The point that Thompson wanted to make was that it wasn't easy to balance the elements in a given ecological system, especially in small, closed environments. Every now and then he'd say there was a lesson here for space colonies, which were roughly analogous to the bottles of pond water. Because of their small size and delicate balance, both of these small ecological units would be difficult to keep alive and healthy over long periods.

Thompson thought that this knowledge would be useful to his students the time that Keith Henson came to the campus. Thompson, who was lecture coordinator on campus, had invited him there to give a talk.

So Keith Henson showed up on this small outpost of civilization and delivered his patented space-colony lecture before an audience of about fifty students and faculty members. He presented them with the standard L5 Society–Gerry O'Neill vision of paradise: the vast orbiting colonies, the lush parklands and mountain ranges rolled up into a closed cylinder, the picture of mommies and daddies rocketing off to work in the space factory while the kiddies attended their classes at Sky High and afterward gamboled somewhere up overhead playing their zero-G space tag.

Thompson was surprised when many of his students fell for this hook, line, and sinker. In fact he was dumbfounded by the semi-religious mania of these kids. He felt like he was at a revival meeting, watching this new-wave space padre up on the podium there, Father Henson, preaching to the converted. It was an awful sight.

"It got me really mad," Thompson said later. "It was just pie-in-the-sky stuff. Henson was selling this concept like it was a real estate development."

The worst of it was that neither Henson, nor Gerard O'Neill, nor anyone else who was pushing space colonization seemed to consider the possible downsides of their meddling with nature. It was as if going into space would be all benefit and no cost—the ultimate universal panacea. In David Thompson's view, this kind of damn-the-consequences thinking was environmentally risky, to say the least.

As an example there was the famous "Maybe-we'll-ignite-the-atmosphere" business back when the Los Alamos scientists were building the A-bomb. Edward Teller had thought about it, the chance that the atomic explosion might light up the surrounding air and that this conflagration would then propagate itself around the world. Some of the bomb makers had even calculated the numerical odds of this actually happening, coming up with the figure of three chances in a million that they'd incinerate the earth. Nevertheless, they went ahead and exploded the bomb.

David Thompson knew from personal experience what happened

when you looked only at benefits and ignored costs. He'd seen it for himself down in that other "last frontier," Antarctica.

Thompson was an ornithologist and had spent three summers in Antarctica observing the behavior of penguins. He'd gone down there and seen with his own eyes that parts of the continent had already been turned into vast junkpiles. There were problems with air pollution, sewage, and waste disposal. The U.S. Navy had once dumped two thousand empty fuel drums into the ocean off Cape Hallett, and many of them washed back up onto the beach, making an unholy mess. So far as Thompson could tell, the same thing would be likely to happen in space.

Not long after Keith Henson's lecture Thompson decided that he'd put together a little space-colony talk of his own, to try to temper some of the Space-Ranger evangelical fervor he'd seen in Henson and in his own students. He wanted these people to see the difference between Henson's claims, which Thompson viewed basically as *advertisements,* and what would likely turn out to be the practical reality of the situation.

It so happened that an apartment complex was going up right next to the campus, and Thompson thought that this would make for another fine object lesson. So he went over to the architect's office and got some of those beautiful before-the-fact architectural renderings of what the place was supposed to look like after it was finished.

Naturally it looked like the Paradise of the Midwest: there were towering trees all over, aesthetically correct clusters of low shrubbery, and lush, grassy acreage—it was beautiful. These renderings looked, in fact, suspiciously of a piece with the paintings that Keith Henson had shown of space colony interiors: parklands, streams, forests, low mountain ranges off in the distance—like the San Francisco Bay Area in miniature. But Thompson was also a published photographer, so he went out to that same apartment complex after it was finished and took a few shots. These revealed an entirely different reality: the promised trees were never planted, there was bare dirt and erosion where the lawns were supposed to be, there were rainwater gullies at several places, and so on.

"You're being sold a real estate development!" he told his audiences. "A space colony's probably going to look more like the inside of a Greyhound bus than it will look like any of these paintings."

After a while he got the idea for an article about it, about the probable consequences of a major move out into space. He'd show what already happened up there in orbit, and what was likely to happen in the future if people didn't pay more attention to such details as environmental impact and ecological costs. So he went off on a research program, even going to NASA headquarters in Washington for data and photographs.

His results were published in the summer 1978 issue of Stewart Brand's journal, *CoEvolution Quarterly*. Entitled "Astropollution," it was a landmark piece. It documented the way in which outer space had already been turned into a garbage dump. "By March 12, 1978," he wrote, "there were 4,078 objects in earth orbit, including 547 payloads and 3,531 trackable pieces of debris."

Some of this space junk, the author showed, had a disconcerting way of reentering the atmosphere where it was least wanted. To illustrate the point he ran photos of charred rocket parts that had fallen in Cape Province, South Africa; Marietta, Ohio; and Winter Haven, Florida. One piece had come down in Cuba, killing a cow.

And the illustrations, too, were a little different from the ones normally seen by the greater American viewing public. There was the shot taken by one of the *Apollo 17* astronauts, for example, which pictured what amounted to a trash heap on the moon—discarded wires, bent pieces of metal, broken parts, abandoned instruments, and you could hardly tell what else. According to NASA, these were the remains of ALSEP, the Apollo Lunar Satellite Experiment Package, but it looked more as if the brave Space Rangers had gone up there, spent a few days camping out on the lunar highlands, and then left in a rush without bothering to pick up after themselves.

Man had landed on the moon and in a matter of hours the place was *a dump!* That was the fact of the matter, and there could only be worse in the offing, as humanity pushed ever farther out into space. Thompson noted how Gerard O'Neill had advocated moving asteroids around by using mass-drivers for propulsion: supposedly these thrusters would work by ejecting chunks of asteroid out the rear end, moving the thing forward by the principle of action-reaction. The unseen reality here, though, was that each and every one of those ejected asteroid chunks would become another piece of floating space garbage. Not only that, these fragments would be hazards to anything in their path: they'd fly through the solar system like stone cannonballs, laying waste to anything they met up with. Somehow, this was another small point that got Lost in the Mania.

As far as David Thompson could see, this business of conquering the solar system was going to have a lot more unwanted side effects than appeared at first glance. He wondered whether the benefits would be worth it.

*F*or Freeman Dyson, the main reason why you had to enclose the solar system was to make room for a growing population. "Malthusian pressures," Dyson had said, "will ultimately drive an intelligent species to adopt some such efficient exploitation of its available resources. One should expect that, within a few thousand years of its entering the stage of industrial development, any intelligent species should be found occupying an artificial biosphere which completely surrounds its parent star."

And for Dave Criswell, too, the essence of life was constant growth and proliferation. "Once you get growth started," he said, "it's not obvious what can stop it." Which, to Criswell, was just as it should be. The universe, after all, was just *dead matter,* and the more of it that got converted into life and mind the better.

Others, such as Frank Tipler, were even more explicit about the need for life to keep expanding out into the universe until the cosmos had been completely subdued. "If life is to survive at all, at just the bare subsistence level," he said, "it must necessarily engulf the entire universe at some point in the future. It does not have the option of remaining in a limited region. Mere survival dictates expansion."

So the whole motive behind the space-colonization scenario was to prevent overpopulation problems here on earth. Now an *engineer's* idea of overpopulation differed fundamentally from that of less hubristic thinkers, many of whom—such as the Club of Rome people—were convinced that with a population of merely four or five billion the earth had already reached, if not exceeded, its carrying capacity. But according to the more progressive, this was sheer nonsense. Not only Bob Truax—who once calculated that the earth could support "a total population of about a septillion" (10^{24}) if they were packed together densely enough, in skyscrapers and so on—but even mainstream Harvard University social scientist types were saying that the carrying capacity of planet Earth had not been even remotely approached.

In 1985 Harvey Brooks, of the John F. Kennedy School of Government at Harvard, claimed that "the world could support a population of a trillion (10^{12}) people at a material standard of living better than that of the most affluent countries." This would require some new living space—"two-thirds of the human population would inhabit artificial islands in the world's oceans," he said—it would require some unconventional farming techniques, and much else besides, but with the right technologies the goal could certainly be achieved.

"None of this would involve implausible extrapolation from potential scientific and technological capabilities we can identify in the laboratory today. It would not violate any fundamental or biological principles."

In making these claims Brooks was relying on the work of Cesare Marchetti, a forward-looking physicist working at the International Institute for Applied Systems Analysis, in Laxenburg, Austria. The center was a fabulous place, located in the former summer palace of Queen Maria Theresa, and Marchetti was its most audacious thinker.

Indeed, Marchetti stood so far back from the ordinary ebb and flow of academic publishing that he'd formulated a set of "fashion wave equations," as he called them, to measure the output of all the other, lower-echelon academics who, whenever a new subject arose, immediately burst forth with "a huge amount of recycled paper."

In 1988 Marchetti had studied the literature on carbon dioxide buildup in the atmosphere, a topic about which even those not so jaundiced as he was had to admit there'd been a lot of dull repetition. But Marchetti found, through his "fashion wave equations," that a total of approximately fourteen hundred papers would be published before these writers had finally exhausted themselves.

"The point of maximum production of papers," he said, "was passed already in 1984 and we are now on the ebb side of the wave. The time constant being seventeen years means [that] in approximately 1992 90 percent of the papers on the subject will have been written."

Marchetti had written about a hundred or so papers of his own, and had published one in 1979, called "10^{12}: A Check on the Earth Carrying Capacity for Man." This came out in the American technical journal *Energy*, where he argued that, "from a technological point of view, a trillion people can live beautifully on the earth, for an unlimited time and without exhausting any primary resource and without overloading the environment."

All you needed to accomplish these miracles, he said, were science, technology, and a little well-placed geoengineering. As in Bob Truax's scenario, people in such a world would be building upward instead of being sprawled all over the countryside. Nevertheless, their habitats would be built on the human scale, and would resemble medieval cities more than they'd resemble skyscrapers. "These cities, like the Amazon rain forest," Marchetti claimed, "will be essentially closed systems where most of the materials, including water, will be recycled, the only physical input being free energy and the only output heat." To get rid of the excess heat, the earth's reflectivity, or albedo, would be modified, "a very easy operation with a sizable fraction of the earth's surface built up."

The population of a trillion would be reached by disconnecting people from nature to the extent that man's "coupling with the earth will be practically nil." You'd employ agricultural techniques, for example, that went far beyond the kind of high-density farming advocated by the Hensons for space colonies. In fact, what he envisioned was more like Eric Drexler's plan for synthesizing food by programmed molecular machinery. People would eat "biosynthetic food" produced by special-purpose microorganisms.

"Some conventional agriculture can be kept," Marchetti allowed, "for the aesthetic enjoyment of flowers and wines."

With the appropriate new technologies and geological adjustments, therefore, the earth could support a trillion people in relative comfort. Even so, sooner or later the planet would become so crowded and used up that space colonization would become a practical necessity. In fact, if population growth kept on the way it always had, then eventually the *other* planets would be used up too. *Then* what would the human race do for its resources and raw materials?

This was just the question on Dave Criswell's mind as he read an article in *Science* magazine. It was in 1976, at about the time Criswell started working for Cal Space, part of the University of California, when he read "The Age of Substitutability," by H. E. Goeller and Alvin M. Weinberg, a masterful summary of humanity's current and probable future use of natural resources. Billing themselves as "cornucopians" (as opposed to the "catastrophists" who had produced the Club of Rome report), Goeller and Weinberg argued that "most of the [earth's] essential raw materials are in infinite supply: that as society exhausts one raw material, it will turn to lower-grade inexhaustible substitutes."

Seemingly, this showed that there would never be any problem with raw materials: if resources were in fact "inexhaustible," then how could there be? But on closer inspection Criswell saw that the authors were talking about a world population that held roughly *stable,* which in Criswell's view was an absolutely unrealistic assumption. The earth's population had increased since day one, was still increasing, and would probably continue to do so for a long time into the future. And if this was true then sooner or later earthly raw materials would in fact run out.

The crux of the matter lay in a statistic that Criswell fastened on and mulled over for a long time. Goeller and Weinberg had computed a numerical value for the total amount of raw materials of all types that were currently being processed worldwide, commodities such as sand and gravel, coal and oil, and metal ores. The figure was 17.3 billion tons per annum: that was the amount of basic raw material that the human race was using per year, circa 1970.

Reading further, Criswell found that for the past four hundred years materials processing had grown at about a 6 percent annual rate. This figure was amazingly steady and reliable, and it represented a twelve-year doubling period. Roughly every twelve years since the 1600s there had been a doubling in the amount of raw materials that had been mined, processed, and used.

But that was only for *earthly* mass handling. Once people started living in space, Criswell thought, materials processing would go through the roof. You might see a rise from the old 6 percent rate to a *20* percent–a-year level, a figure that would give you a doubling period of only 3.6 years. At that rate you'd be gobbling up the extant solar system quickly: the asteroids would be gone by A.D. 2140, and the planet Jupiter could conceivably have vanished by the year 2600.

"Once you got space industry going," Criswell said later, "it wouldn't really matter if it grew at 6 percent a year or 20 percent, because in either case that growth would soon eat up the minor objects revolving around the sun. In other words, once things really got going *the solar system itself* would turn out to be a trivial resource."

And all of a sudden it seemed to Dave Criswell that the great human expansion would have to stop after all. The solar system was finite, and when you ran out of it, that would be the end of that.

Or was it? Criswell had a Ph.D. in space physics and astronomy from Rice University and knew that there was a constant infall of new matter into the solar system. Comets kept coming in from the Oort cloud, and if enough of these streamed in then maybe the great human expansion could continue.

At this point—it was sometime in 1979—Criswell read another article, this one in *Scientific American,* which gave the latest estimate for the rate of new cometary infall. Unfortunately, the rate was only on the order of some fifteen or so new objects every million years. This figure was *so* small, it amounted to *so* much less than the quantity of matter that mankind was processing on earth already, that he finally thought the situation was hopeless.

A few years later, though, Dave Criswell had an amazing thought. He suddenly realized that even after all the planets had been used up, the greatest resource in the solar system would still remain intact. Indeed, it would be right there overhead—the sun itself, a body so massive that it contained *fully 99 percent of all the matter in the solar system.*

Nobody had ever thought of regarding the sun as a "natural resource" before, but now Dave Criswell would.

*T*hat was on the hubristic side, even a little demented, perhaps. But on the other hand, was the idea of dismantling the sun anything more than a large-scale extension of the schemes that had been proposed for the natural features of earth? Some of these schemes were, admittedly, a little oddball—as, for example, the various things people wanted to do with icebergs.

In World War II a member of Lord Mountbatten's staff, Geoffrey Pike, came up with an idea for turning icebergs into aircraft carriers. His idea was to take an iceberg—one that was about a half-mile long, more or less—surround it with refrigeration coils to keep it from melting, and send it out on the high seas. Whatever else could happen to it, the iceberg aircraft carrier at least *couldn't sink,* and this made the scheme attractive enough to the political leaders at the time that even Winston Churchill approved of the idea: "Let us cut a large chunk of ice from the Arctic ice cap and tow it down past Cornwall, fly on our aircraft and tow it to the point of attack," he said. (During World War I Will Rogers had proposed boiling the Atlantic Ocean to clear it of U-boats, an idea that not even Churchill took seriously.)

A prototype iceberg aircraft carrier was built at Patricia Lake near Jasper, Canada, using a block of ice sixty feet long, thirty feet wide, twenty feet deep, and weighing a thousand tons. This test model was kept frozen for about six months, from the winter of 1942–3 to the end of the following summer. Shortly afterward, though, people started having second thoughts about iceberg aircraft carriers, and soon enough the project was, as Churchill later reported, "reluctantly dropped."

An even earlier idea was to tow icebergs to Africa and melt them down for their water. That would not only irrigate the deserts that most needed it, it would have the added benefit of getting icebergs out of shipping lanes. The scheme had been proposed time and again by a wide variety of advanced thinkers starting with Erasmus

Darwin in the 1700s, and finally, in 1974, an "Iceberg Transport Seminar" was held at the Institute on Man and Science, in Rensselaerville, New York. The seminar was four days long, but in the end most of the plans for making good use of icebergs were deemed not viable. "After looking at the practicalities involved—water depths at delivery points, markets able to pay for the water, legal complications, etc.—the idea seems to have died a quiet death," one of the conferees reported, sadly. "Uncaught, the icebergs simply melt away."

That was the closest icebergs ever got to doing useful work, but the idea of irrigating Africa with foreign water had several later reincarnations. In 1975 Joseph Debanné, a professor at the University of Ottawa, proposed that Algeria buy the Rhône River from France and send its water coursing across the Mediterranean in a big plastic pipe. Nothing ever came of this idea either, but not long afterward Frank Davidson, head of the Macro-Engineering Research Group at MIT, went the professor one better. Davidson suggested running an enormous tube from the Amazon River, in South America, to the Sahara Desert, in Africa. This was a long way, across the whole southern Atlantic, in fact, but so what? Hadn't the transatlantic cable already been laid in the 1860s? All you needed in the present case was a way of keeping the water tube from floating to the surface, and Davidson had this problem solved too. "An aqueduct 150 feet in diameter, made of rubber or rubberlike plastic, could be held in place on the sea bottom by a cement ballast to overcome the buoyancy of the seawater," he said.

It was a little off-center, this plan, but it was not so bad once you actually looked at a map and saw that the Amazon River was in fact *pointed right at Africa,* as if that's where all that water—some seventeen billion gallons per hour, on average—was actually meant to end up.

The fact was that macroengineers had thrown around lots of ideas bolder than merely changing the course of mighty rivers such as the Amazon. There was a plan for damming up the Bering Straits to confine the coldest waters of the Arctic, thereby preventing a new ice age; there was a proposal for walling off the Strait of Gibraltar so that some of the Mediterranean Sea could be drained, providing Europeans with more living space. These hubristic schemes, of course, had to be seen in perspective, which was provided by the fact that some of them had been tried, and had even *worked*. About half the population of Holland lived on land that had been reclaimed from the North Sea. And what was the Panama Canal but a river that man had created for the express purpose of separating two continents that nature had (rather thoughtlessly) joined together? No one gave these things a thought *after* they'd been done, but *beforehand*, of course, the ideas were "crazy."

So why not the icebergs-to-Africa tactic? Or Davidson's Amazon-to-the-Sahara scenario? Or even Dave Criswell's dismantling-the-sun project, for that matter?

Criswell's visions of a space-based civilization went back to his childhood days in Texas. "I can remember back to the age of six or seven," he recalled, "lying in bed, in my granddad's house. I can remember waking up late at night, when the windows were open. The windows had lace curtains on them and the wind would move the curtains apart, and then I could see the moon coming up.

"How peaceful it looked. I suspect I got a keen attachment to the moon from looking at it and from wondering about what it would be like to go up there."

The first book he ever read on his own had a chapter in it about a trip to the moon. Dave was in the fourth grade at the time, about ten years old. The book said that the moon had no air, that conditions were harsh, that the day-night cycle was a month long. It also had a picture of the spaceship that would make the flight.

"The spaceship looked like something out of the old Buck Rogers movies: bulbous nose, fins; it landed horizontally. The moon was shown as having a rocky, mountain-type appearance, which is nothing at all like it really is. On the moon, the actual features are very subdued, beaten-down by all the sandblasting from meteorites."

And then there were the *Collier's* magazine articles. These had been published back in the early fifties, vivid descriptions of space stations, voyages to the moon, exploring Mars, and so on. They'd been written by Wernher von Braun and Willy Ley and were illustrated with the world's most romantic space art, by Chesley

Bonestell. This was also the period of "Tom Corbett, Space Cadet," "Buck Rogers," and other space shows on television. Going up into outer space, Criswell soon came to feel, would be a necessary element of his life.

Later the family moved to Los Alamos, where his father, a chemist, had gotten a job.

"Los Alamos was a mysterious place. I remember hearing these explosions down in the finger canyons. There were warning signs all over not to go down into the canyons, but they were still doing experiments down there with A-bomb detonators, and every once in a while you'd be standing there looking through the fence and *Boom!* It would shake the ground right under you."

After he got his degree he returned to Los Alamos every so often as a consultant to the lab, and this is how he got to know Eric Jones, who was a senior scientist there. One time Jones and an anthropologist by the name of Ben Finney were organizing a conference on interstellar migration to be held at the lab, and Jones invited Criswell. The question of solar system resources was so serious to Dave Criswell that he took a month off from his regular work, without pay, for the sole purpose of trying to figure out how you'd keep the human race expanding once the planets ran out.

Criswell had an office at the California Space Institute, which was located on a cliff overlooking the Pacific in La Jolla. His office was in a little white clapboard house right at the edge of the cliff, smack up against the ocean, so he'd come in early in the morning, sit down at his desk, and stare off at the blue Pacific. He liked the sight of the sunlight glinting off the water; it was somewhat hypnotic. Every once in a while he'd see a couple of dolphins jumping in and out of the waves; less often a whale might surface, and occasionally even a nuclear submarine.

One morning the sun was coming up over the hills and as it shined out over the water Dave Criswell contemplated the amount of solar stuff that was radiating away off into the void—all those high-quality photons and valuable solar-element abundances, all of them winging their way off to nowhere and being lost forever—when suddenly the errant thought struck him: *Gee, how many Pacific Oceans could you make out of that lost solar stuff?*

From that moment on Criswell's mind worked at an entirely

different scale. The sun was *raw material*. The question was how to gain access to it.

"So I just sat back and thought to myself: *Well, how could you do it? What would work?* And I came up with two extremely crude, seat-of-the-pants methods. It probably took me about ten minutes to come up with them, but what I thought was, you can squeeze the sun or you can spin it. Either way you'd be able to get some of its matter out."

You could squeeze the sun by circling it with a ring of some kind, then tightening the ring, forcing some solar matter up and off the surface. But a ring of *what*? Well, how about a gigantic circle of particle accelerators? You could construct it in sections so that you'd have a collection of pipe segments going around the sun horizontally at the equator. From far away it might look like a dashed line going completely around the sun. You could fire particle beams through those lined-up sections, creating magnetic currents that would draw them in toward each other, tightening and loosening the belt in pulsating waves. It would be like a rubber band squeezing the sun at its equator and pushing solar plasma off toward the sun's poles, where it would fly off in spurts, as if through a rocket nozzle. As Criswell would come to think of it, this was "the huff-and-puff method" of lifting solar matter away from the surface.

So there was Texas Dave Criswell sitting in his tiny wood house by the sea there in La Jolla, squinting off into the ocean haze and thinking about huffing and puffing the sun apart. The situation didn't strike him as abnormal in the least, but on the other hand he didn't go running across campus telling all and sundry about this new idea of his for saving humanity. Not that there were many other people around: the only other person working in the house on the cliff was Catherine Gautier, another physicist who later became associate director of Cal Space. They'd say hello to each other in the morning and chat over coffee, but that was about the extent of it.

Huff 'n' puff, that was one method. The other one involved speeding up the sun's natural rotation rate until some of its surface materials were slung off by centrifugal force. You could accomplish this with accelerators, too, although you'd have to place them in a different orbital plane. Instead of placing them horizontally around the equator you'd line them up vertically, so that they went around the sun's two poles like lines of longitude. And then if you started that whole plane of accelerators spinning fast (say, by attaching rockets to each of the accelerator units), they'd sooner or later start pulling the sun along with them. The whole system, the sun and its vertical plane of accelerators, would spin faster and faster until some of that solar plasma broke off from the sun's surface and came flying out into space.

And that was Criswell's second method for taking the sun apart.

As he added in various details, Criswell gradually came to realize that *star lifting*, as he called it, would be a godsend to humanity. Not only would it provide for a virtually endless supply of matter, it would, as an added benefit, lengthen the lifetime of Mother Sun herself. It was a well-known fact of astrophysics that some types of stars had longer lifetimes than others. Our very own sun was a yellow star and had a certain definite lifetime. But if you whittled the sun down a bit you might be able to turn it into a type of star that had a longer expected life span. So you could also think of star lifting as a process of *stellar husbandry*, or taking care of your star.

You could also store the lifted solar matter and then, when the time came, put it back together again in the form of a man-made sun, an *industrial star*. By artful handling of the original solar core you might even be able to convert the sun's center into something else, a *cultured black hole*. In fact, there was virtually no limit to what you could accomplish once you took the long view, thought on the right scale, and began to see the celestial objects of the universe for what they really were, which was to say, gobs of brute matter awaiting intelligent transformation.

David Thompson, as might be expected, was not an unqualified fan of Criswell's dismantling-the-sun project.

"Scientists like to push back the limits, and that's a good thing," Thompson said. "It lets you know how far you can go, and Criswell is taking the engineering approach to its limits.

"But these things can get unbalanced. When you get to *extreme* limits, a moral component has to enter in. Some people like to push the social limits: you know, they want to see how many women they can rape and murder. Or they want to see what it feels like to carve up Grandma. So before you start carving up the sun you've got to think about it from a moral perspective. The human species doesn't necessarily have sole ownership of the earth, the sun, or the solar system. You've got to consider how any given plan will affect the other species on the planet, and maybe even other possible civilizations in the Galaxy. What if alien civilizations are using the sun as their North Star, for example? They might not like to see it dimmed."

Keith Henson, on the other hand, saw nothing but salvation in Criswell's stellar husbandry plans. "You want to take good care of your star," he said, "otherwise it gets all dark and icky."

But what if the worst happened? What if Criswell's star lifting went a little too far and Old Sol actually . . . *blinked out?* Or the lifted solar matter got a bit out of control and . . . accidentally set fire to the earth and the moon, burning them both to a crisp? If it went the least bit wrong, a plan like Criswell's could turn the whole solar system into a garbage heap.

Sooner or later, of course, the solar system would turn to garbage anyway, even if we left the sun alone. Indeed it was precisely this fact, that the solar system was not going to last forever—so you shouldn't put all your eggs in one basket—that led others to advocate colonizing the stars.

For a long time interstellar travel was regarded as the most advanced of all high-tech projects. But some scientists, especially those of the more close-to-nature variety, were soon finding ways of traveling to the stars without using rockets of any sort. There were entirely *natural* ways of getting out there, ones that made no use of faster-than-light rockets, warp drives, or any other science-fictional propulsion devices. According to Eric Jones and Ben Finney, you could hitch a ride on a comet.

Ben Finney, the son of a navy pilot, was an anthropologist at the University of Hawaii. He'd spent many long hours aboard surfboards in his youth, and later even coauthored two books on the subject, *Surfing, the Sport of Hawaiian Kings,* and *A Pictorial History of Surfing,* plus a large clutch of related scholarly articles.

Although he grew up in San Diego, he'd sailed and surfed all over the Pacific, and when he got his Ph.D. from Harvard it was for a dissertation that he wrote about Tahiti, where he'd lived for a while. Later Finney became an expert in Polynesian ocean voyaging, unearthing all sorts of little-known lore about silent paddling, outrigger canoes, ancient navigational techniques, and so forth. The more he studied the Polynesians the more respect he acquired for their sailing abilities, and in fact it was their great seamanship that got him to thinking about Thor Heyerdahl's famous trip aboard the raft *Kon-Tiki*.

Heyerdahl had wanted to prove out his own theory about where the Polynesians had come from. The linguistic evidence seemed to show that they'd originated somewhere in Southeast Asia, from the area of Indonesia and the Philippines, but Heyerdahl pointed out that this meant the Polynesians would have had to travel east, against the prevailing winds and currents, which was a feat he thought them incapable of. He therefore proposed a countertheory of his own, that the Polynesians must have sailed westward from the coast of South America. To establish that they could have done it that way, Heyerdahl built the *Kon-Tiki* and rode it from Peru to the Tuamotu Islands, taking 101 days to make the trip. As for the linguistic evidence, Heyerdahl argued that some Indonesians could have gotten to the South Pacific by following the currents north past Japan, eastward to the coast of North America, and then finally downward to the south.

Finney didn't buy this reasoning at all. The Polynesians, he thought, were some of the best ocean sailors of all time; how could they not have had the ability to tack against the winds? And if they *could* sail against wind and current then they could easily have made the trip from the Southeast Asian waters eastward across the Pacific, and finally to the isolated outposts of Hawaii and Easter Island.

What Finney wanted to prove was the direct opposite of what Heyerdahl had claimed, but he wanted to prove it in the same way, with an experimental voyage in an ancient-style watercraft. So he founded the Polynesian Voyaging Society to build such a craft. It would be named *Hokule'a*, Hawaiian for "Arcturus," a star that passed directly over the Hawaiian Islands.

Hokule'a was intended to duplicate the kind of ship the Polynesians used six hundred to one thousand years ago, during the time of the great ocean crossings, and it was essentially two canoe hulls lashed together in parallel with a deck running across the tops, much in the manner of a latter-day catamaran such as the Hobie Cat. It had sails fore and midships, was sixty feet long and seventeen feet wide, and could easily hold twenty people.

Together with a crew of seventeen, many of whom were native Hawaiians, Finney set out from Honolulu bound for Tahiti on May 1, 1976. The voyagers took with them all the essential elements of a small founding population: some animals—dogs, pigs, and chickens—plus an assortment of plants and seeds. They had plenty of food and water, and the only items of consequence they lacked were navigational equipment. This was intentional: the ancient voyagers sailed without benefit of charts, compass, sextant, and so would Ben Finney and friends. They'd find their way across three thousand miles of open water only by reference to the forces of nature: the wind, the waves, cloud formations, and the sun, moon, and stars.

They had the usual luck of a sailing vessel: they met with storms, they were becalmed, and sometimes, when the sky was blotted out by clouds, they weren't exactly sure which way they were going.

"One time when I was steering it happened that the bow stars were no good because they were setting," Finney recalled later. "And the stern stars weren't very good either because it was cloudy, so I was using the Southern Cross to steer by—specifically, I was using one of the pointers to the Southern Cross, Proxima Centauri, the closest star to earth. But then the clouds covered that up, too, and when this happens you've got to quickly use stars elsewhere, to get your bearings. So for a while I was steering by the Magellanic Clouds, and I thought to myself, *This is crazy. Here I am on a canoe, sailing through Polynesia without instruments, steering first on the closest star, then the closest galaxy.* It was a little weird."

Anyway, thirty-three days after setting out they landed in Tahiti. The whole experience was a lesson in migration over vast uncharted distances, and the parallel to interstellar travel was not lost on Ben Finney.

He'd always been interested in flying and space travel, and had worked for General Dynamics for a while, back in the States. After the Tahiti trip he put himself back in touch with the space movement, attending various space conferences, incuding one in Houston, where he met Carolyn Henson, who at the time happened to be breast-feeding her baby.

"That was pretty gutsy at an aerospace conference, which is such a male place," Finney recalled. He joined the L5 Society, and later cofounded an L5 chapter at the University of Hawaii.

Finney went to O'Neill's space manufacturing conferences at Princeton, and gave a paper at one of them called "Exploring and Settling Pacific Ocean Space—Past Analogues for Future Events?" In the audience was Eric Jones who, like many space fans, was a Polynesian buff.

Jones, committed L5 member that he was, had a special interest in getting humanity pointed toward the stars. In fact he'd worked out his own original scheme for doing this: it involved departing the solar system by hitching rides aboard interstellar comets. In its own way, he thought, it would be like the migration of the ancient Polynesians: small founding populations setting off across the ocean to colonize every last speck of land they found.

Jones had gotten the comet-travel idea from reading Freeman Dyson's book *Disturbing the Universe*. Dyson had told how small bands of colonizers could climb aboard a comet, plant a few trees and vegetables on it, and live happily ever after. Comets were largely made of water, and except for the minor inconveniences of lacking an atmosphere or gravity—both of which could be created artificially—they'd make ideal homes for the right type of rugged individualist. "Comets, not planets," Dyson had written, "may be the major potential habitat of life in the solar system."

Jones and Finney now appropriated Dyson's comet idea as a means of making it out to the stars. Here Jones took advantage of recent theories about possible *interstellar* comets.

For a long time it had been thought that all of the solar system's comets remained in orbits around the sun, but then astronomers realized that some of them might in fact escape from the solar system and wander all over the Galaxy. These interstellar comets,

Eric Jones now guessed, would make excellent vehicles for a low-tech human migration wave to Alpha Centauri and points beyond. You could climb aboard an interstellar comet and ride it to the stars for free.

When Jones told Finney about his comet-travel idea Finney saw immediately that it had possibilities. After all, he had experience in migrating across trackless voids, and he could see that the scheme, as eccentric as it seemed at first, just might work.

The cometary travelers, Jones and Finney now claimed, would be nomads, sailing through the void at an extremely slow rate of speed. Interstellar comets moved through space so slowly, in fact, that they wouldn't be likely to reach another star system for at least a hundred thousand years, maybe more. This was a long time, but the great human migrations of the past hadn't exactly been accomplished overnight.

Indeed there were so many parallels between these cometary wanderers and the ancient ones that Jones and Finney began to speak of the nomads as living in "bands" or "tribes," and compared them to the Australian aborigines and to the !Kung peoples of the Kalahari Desert. The nomads, after all, would travel through space in small groups of about twenty-five or so, which was about the size of the aboriginal hunter-gatherer bands. To prevent inbreeding, the ancient tribes had gathered together in larger communities of about five hundred, and so too would the space travelers.

"We expect that cometary bands would also cluster for purposes of healthy breeding and social enrichment in tribal groups of at least five hundred citizens. We imagine clusters of some twenty cometary bands, the members of which would exchange marriageable youths and periodically meet to celebrate rites of passage, for calendrical observances, and for other social occasions."

It was the low road to Proxima Centauri.

*T*en years after writing "Astropollution," David Thompson had not really changed his mind about space colonies except that he increasingly saw them as inevitable. They weren't even inherently *bad* unless you thought of them as a panacea, as a technological fix that would allow the human race to escape its basic problems of pollution, war, and overpopulation. Space colonies were not a refuge from overdevelopment, because they represented still more development, only "off in space."

On the other hand Thompson *did* change his mind about "limits." What caused him to see things differently was the failure of the Club of Rome's predictions to come true.

"The Club of Rome seemed to prove that there were limits, and that we were already close to them, or beyond," he said. "But now that the Club's predictions have not been met, I have to admit that if there are limits, we cannot predict where they lie. What the Club of Rome people forgot was that as you produce more and more people you also get more and more *brains*. You get more scientists who can figure out different ways to attack problems. If you double the population you get twice as many scientists."

And if there *were* limits, it was only natural, biologically speaking, to try to go beyond them.

"In a strict biological sense," Thompson said, "our pushing the limits isn't good or bad, it's just what species do. Every species pushes its territorial limits and tries to increase its population if possible; the human species is no different in trying to do the same. Growth is inherently protective because the more areas are settled by a species the safer that species is from natural catastrophes, epidemics, and so forth. What's different about the human species is that we have such tremendous power to override the limits, to overwhelm and destroy the planet, while at the same time we also have the foresight to avoid it. Perhaps by colonizing space and nearby planetary systems, we're just extending the range of our species, providing new opportunities for speciation, and ensuring our survival against the death of our planet or the sun.

"But now comes the moral statement: I hope we can do this without wrecking the home planet."

Space colonization had been opposed by lots of environmentalists as fostering a "disposable planet mentality," and anyone had to admit that this was a criticism of some justice, given the way Freeman Dyson had imagined turning Jupiter into a tidy solar system enclosure, the way Dave Criswell imagined using the planet Mercury to make the accelerators needed for his stellar husbandry project, and so on. Some environmentalists and social critics, seeing space colonization as inevitable over the long run, finally came out with the idea of setting aside certain precincts of the solar system—or the universe at large—as "solar system wilderness areas," or "space preserves." Tranquillity Base on the moon, where *Apollo 11* landed, ought to be set aside, as should other such historical sites.

No one could possibly object to that, but then there were arguments over other cases, such as the rings of Saturn. If they contained valuable ices and minerals, should they be mined to the point of invisibility? Proponents said yes, arguing that very few human beings had ever actually *seen* Saturn's rings—to which opponents replied that that didn't matter: it was good enough merely knowing the rings were still there.

The ecological disputes of the future—some of them—would not be about whales but about *saving Saturn's rings*. Still, it was largely a matter of perspective. Hans Moravec had said of ordinary matter that it didn't *do* anything when it was left to its own devices—that was why we had to step in and make improvements. Humanity was the universe's way of transforming itself into something higher than countless blobs of inert mass.

Dave Criswell saw things in much the same terms. He didn't so much want to *dispose* of the planets as he wanted to *convert them into life* . . . or better yet into *mind itself.*

"The question that got me to thinking about all this was very simple: *What fraction of the universe can you turn into mind?*" (Frank Tipler had his own answer to that. "When life has completely engulfed the entire universe," he said, "it will incorporate more and more material into itself, and the distinction between living and nonliving matter will lose its meaning.")

Criswell used to wonder what the night sky would look like after most of the stars had been husbanded, whether by ourselves, our successors, or even by other civilizations. What would the Milky Way look like, he asked himself, after all those points of light had been turned into industrial stars and cultured black holes, after they'd been converted into space habitats and macromachines?

In fact he thought there must be plenty of advanced civilizations out there doing this stuff *already*, which only led to the question of why there were so many stars out there still shining. Why hadn't those advanced civilizations used up *all* the stars? Why wasn't the night sky *totally* black?

There had to be a reason.

"Why leave the stars?" Dave Criswell wondered. "Are they the flower gardens of advanced civilizations? Are galaxies decorative pieces? *What good are stars?*"

8

Death of the Impossible

*A*ccording to Barrow and Tipler in *The Anthropic Cosmological Principle,* when intelligent life reaches the Omega Point then it will have "gained control of *all* matter and forces not only in a single universe, but in all universes whose existence is logically possible; life will have spread into *all* spatial regions in all universes which could logically exist, and will have stored an infinite amount of information, including *all* bits of knowledge which it is logically possible to know."

With people like Eric Drexler giving us complete control over the structure of matter, Hans Moravec making people over into near-omnipotent bush robots, Dave Criswell telling us how to make industrial stars and cultured black holes, and Eric Jones and Ben Finney showing us Mother Nature's own way of traveling to the stars—with all this laid out in front of them, scientists pretty much knew all they needed to know to accomplish the whole Barrow and Tipler program. And as ambitious as it was, there was no reason to think that any of it was in the least impossible. Nothing predicted by Barrow and Tipler seemed to violate the known laws of nature; none of it invoked magic or mysticism. Quite the contrary, everything they envisioned seemed to follow naturally from the normal and ordinary progress of science.

In fact it was hard to think of anything that could *prevent* the complete dominance of mind over matter from eventually taking place. All you'd need to get around any obstacles were ingenuity, time, and energy, and who could imagine that these would be in short supply in the indefinite future? Gerald Feinberg, the Columbia University physicist, once went so far as to declare that "everything possible will eventually be accomplished." He didn't even think it would take very long for this to happen: "I am inclined to put two hundred years as an upper limit for the accomplishment of any possibility that we can imagine today."

Well, that of course left only the *impossible* as the one thing remaining for hubristic intellectual adventurers to whittle away at. Feinberg, for one, thought that they'd succeed even here. "Everything will be accomplished that does not violate known fundamental laws of science," he said, "as well as many things that do violate those laws."

So in no small numbers scientists tried to do the impossible. And how understandable this was. For what does the independent and inquiring mind hate more than being told that *you can't do* something, or that something just *can't be done*, pure and simple, by any agency at all, at any time, no matter what. Indeed, the whole concept of the "impossible" was something of an affront to the human spirit; it was a knock in the ribs, a slap in the face of creativity and advanced intelligence, which was why being told that something was impossible was an unparalleled stimulus for getting all sorts of people to try to accomplish it anyway, as witness all the attempts to build perpetual motion machines, antigravity generators, time-travel vehicles, and all the rest.

Besides, there was always the residual possibility that the naysayers would turn out to be *wrong* and the yea-sayers *right*, and that one day the latter would reappear to laugh in your face. As one cryonicist put it, "When *you* die, you're dead. When *I* die, I might come back. So who's the dummy?"

It was a point worth considering. How many times in the past had certain things been said to be *impossible*, only to have it turn out shortly thereafter that the item in question had already been done or soon would be. What greater cliché was there in the history of science than the comic litany of false it-couldn't-be-dones: the infamous case of Auguste Comte saying, in 1844, that it would never be known what the stars were made of, followed in a few years by the spectroscope being applied to starlight to reveal the stars' chemical composition; or the case of Lord Rutherford, the man who discovered the structure of the atom, saying in 1933 that dreams of controlled nuclear fission were "moonshine."

And those weren't even the *worst* examples. No, the huffiest of all *it-couldn't-be-done* claims centered on the notion that human beings could actually *fly,* either at all, or across long distances, or to the moon, the stars, or wherever else. It was as if for unstated reasons human flight was something that *couldn't be allowed to happen.* "The demonstration that no possible combination of known substances, known forms of machinery and known forms of force, can be united in a practical machine by which man shall fly long distances through the air, seems to the writer as complete as it is possible for the demonstration of any physical fact to be." That was Simon Newcomb, the Johns Hopkins University mathematician and astronomer in 1906, three years *after* the Wright brothers actually flew.

There had been so many embarrassments of this type that at about midcentury Arthur C. Clarke came out with a guideline for avoiding them, which he termed Clarke's Law: "When a distinguished but elderly scientist states that something is possible, he is almost certainly right. When he states that something is impossible, he is very probably wrong."

Still, one had to admit there were lots of things left that were *really and truly impossible,* even if it took some ingenuity in coming up with a proper list of examples. Such as: "A camel cannot pass through the eye of a needle." (Well, unless of course it was a very *large* needle.) Or: "It is impossible for a door to be simultaneously open and closed." (Well, unless of course it was a *revolving* door.)

Indeed, watertight examples of the *really and truly impossible* were so exceptionally hard to come by that paradigm cases turned out to be either trivial or absurd. "I know I will never play the piano like Vladimir Horowitz," offered Milton Rothman, a physicist, "no matter how hard I try." Or: "It is a scientific fact that you can't put a thousand gallons of water into a pint bottle"—this from David Park (also a physicist). Or, from Scott Lankford, a mountaineer: "Everest on roller skates." Or, from Michael Katz, a neuroanatomist: "Perpetual motion bees."

No one could bother trying to overcome *those* impossibilities, but off in the distance loomed some other, more metaphysically profound specimens. They beckoned like the Mount Everests of science: antigravity generators, faster-than-light travel, antimatter propulsion, space warps, time machines. There were physicists aplenty who took a look at these peaks and decided they had to climb them.

*T*here was the case of Bob Forward, for example. A physicist at the Hughes Research Laboratories, in Malibu, California, Forward always wanted to work on things that other people considered *impossible*. "I don't ever remember not having this motivation," he said. "It's built into me. It's one of the reasons I wasn't really popular in high school: I didn't want to do what everybody else was doing, I wanted to do something different. Nuclear physics was the rage when I was in graduate school, and so I said to myself, if everybody's going into that, I'm going somewhere else."

Later, interstellar flight became a particular Bob Forward specialty. He bought a black Datsun 280ZX, and the first thing he did when he got it was rip off all the automotive brightwork—"I didn't like all these 'Zs' and 'Xs' and '280s'"—patch over the holes, and then install on the car his own personalized California license plates: "NTRSTLR." That was the only readable word on his Datsun 280ZX.

On the other hand, Bob Forward was one of the few extremely audacious thinkers who never joined the L5 Society. "L5 was too near-term," he said.

Too near-term?

But the blunt fact of the matter was that the L5 Society's core projects—space stations, orbital factories, lunar colonies, and such—had never been especially challenging items in Bob Forward's eyes. "We already know how to do all that," he said. "It's just a matter of money and politics. I've always utilized my energies trying to do something a little harder."

Unequivocally, Forward was the Bob Truax of the Higher Ideational Content. "I feel quite often like one of these backyard inventors—you know, who goes out and builds contraptions in his garage. Except that the pieces of junk that I use are ideas from the forefront of physics."

Not that other physicists always quite approved of the *uses* to which he put their ideas. What Bob Forward wanted to do with antimatter, to cite just one example, was hardly mentionable in polite company, as became clear at a conference commemorating the discovery of the antiproton.

The gathering was held at the University of California at Berkeley, in the fall of 1985, and drew some of the biggest names in science. Six of the fifteen speakers were Nobel laureates in physics, including Owen Chamberlain and Emilio Segrè, who had first detected antiprotons in 1955.

Antimatter had always been big with science fiction buffs because of what happened when it came in contact with ordinary matter, which is to say that the two materials annihilated each other in fabulous bursts of atomic energy. This property had made antimatter a prime candidate for exotic propulsion systems, especially in interstellar rockets. At any rate, for some reason, probably just to liven up the proceedings, Owen Chamberlain put this question to his fellow physicists: "How many believe that antimatter will be used as a rocket fuel within the next fifty years?"

In the audience a rather portly partner rises from his seat, and as soon as they laid eyes on him, everyone present knew who it was. With the white, flowing hair, white bow tie, white shirt, and white vest—who else in his right mind would wear *a white vest?*— it could be none other than *Bob Forward*, who in his richest basso profundo now intoned, "Give me ten minutes, and I'll *prove* it can be done!"

But the people in the audience were *physicists*, and of course they knew that this *couldn't possibly work*, and in any case they certainly didn't want to hear any reckless lectures about it from the rather threatening figure now looming over them. And so no more was heard on that occasion about antimatter rocketry.

Later, though, out in the hallway, Luis Alvarez, the 1968 Nobel Prize winner in physics, came over to Forward to tell him why it was that his antimatter propulsion scheme (whatever it was) *couldn't possibly work.*

"It's crazy," said Alvarez. "The extra shielding you'd need to protect against the gamma rays would erase your fuel savings."

Forward said: "I'm an engineer, and I say it would work."

Alvarez said: "So am I, and I'm known for supporting some pretty far-out ideas. But this is nonsense."

That was the kind of reception Bob Forward got at some of the more mainstream physics conferences. At gatherings of another type, though, he tended to be far more successful. In 1974, about ten years before the antimatter episode, Forward gave a lecture at a meeting of the Science Fiction Writers of America, in Los Angeles, on the topic "Far-Out Physics." He took the opportunity to outline six separate ways of producing antigravity, three different types of time machines, a fifth-dimensional hypervelocity space drive ("This is speculative," he admitted), not even to mention all the standard stuff about black holes, tachyon tunneling, and space warps ("They'd allow point-to-point travel *within* our universe without having to go *through* it," he said).

And the audience *loved* it! Nobody *there* told him he was crazy!

And why should they? As far as they could see, everything that Forward told them was perfectly consistent with the laws of nature. All of it could actually happen, and much of it probably would.

"These things are not really impossible," Bob Forward said. "Just expensive and difficult."

*E*ven after we'd gone to the moon and back, some physicists of a more conservative stripe were still repeating the old skeptical adage "Space travel is bilge," a sentiment first espoused by England's Astronomer Royal, Sir Richard Wooley, in 1956. Wooley's statement was repeated again, word for word, as recently as 1987 by David Park, the distinguished theoretical physicist, in the book *No Way: The Nature of the Impossible,* in which a bunch of scientific experts got together to say what *could* and what *couldn't possibly happen* in the indefinite future.

Park was the Webster Atwell Professor of Physics at Williams College, in Massachusetts, and what he meant by "space travel" when he said it was "bilge" was, specifically, manned, round-trip interstellar flight, something that no one (except for Eric Jones and Ben Finney with their comet-travel scheme) ever claimed would be *easy*. Nevertheless, Park now argued that owing to the fact that chemical rockets used up lots of fuel, interstellar flight was forbidden by the laws of nature, and always would be, forevermore and in perpetuity.

Well, this type of argument was not exactly news to the greater world of Space Rangers. To the contrary, everyone who had ever been involved with rocketry knew full well that, as Carolyn Henson once put it, "chemical rockets just barely work." The reason they just barely worked was that in order to get off the ground such a rocket had to lift not only the payload but also *itself*—meaning all that heavy chemical fuel, the fuel tanks, the rocket motor, and so forth—off the surface. The payload was just *the least little thing*, something that sat at the very tip of the rocket, like a pimple.

Many skeptics, however, saw this fact as constituting such a formidable barrier to rocket flight that they imagined long-distance travel by rocket *could never possibly happen* and offered all sorts of proofs to this effect. In 1926, for example, a British professor by the name of A. W. Bickerton claimed to prove that chemical rockets could never even escape the earth's gravitational field, let alone get beyond it. "This foolish idea of shooting at the moon," Bickerton wrote, "is an example of the absurd length to which vicious specialization will carry scientists working in thought-tight compartments." And in the same vein, Dr. Lee De Forest, inventor of the vacuum tube, predicted only a few months before *Sputnik* that man would never reach the moon, "regardless of all future scientific advances."

So when, thirty years later, David Park came out and said essentially the same thing with respect to interstellar travel, he was placing himself squarely within the confines of a grand and glorious tradition. How fitting it was, then, that Park's argument should turn out to be virtually identical to A. W. Bickerton's. For Park now contended that when you took the interstellar rocket and subtracted (1) the fuel used to launch it toward the stars, (2) the fuel used to halt the rocket once it reached the other star system, and (3) the fuel used to launch it back toward earth, what you had at the end was *the least little thing* you could possibly imagine, like a pimple.

"When the ship finally docks on earth," Park said, "it will have a mass only 0.00000625 as great as it had at departure. If it originally weighed 10,000 tons, it will now weigh 125 pounds. . . . But why pursue the fantasy any further?"

Indeed. By any measure, this was extremely puzzling reasoning. For one thing, an initial lift-off weight of ten thousand tons was not all that stupendous even by NASA standards: the *Saturn V* rockets used for Apollo missions, which after all only went to the moon and back, weighed in at three thousand tons. Bob Truax's *Sea Dragon* rocket, which he'd proposed at Aerojet General back in the 1960s, was supposed to weigh twenty thousand tons at lift-off, fully *twice* the value Park mentioned for flying to the stars, but all the *Sea Dragon* would do was fly up to earth orbit, jettison its cargo, and flop back down again into the sea. Park's argument showed, if anything, that interstellar flight was *entirely* possible so long as you started off with a really gargantuan launch vehicle.

The other limitation of Park's reasoning was that it applied only to *chemical* rocketry, whereas a whole rash of other propulsion systems had been advocated by advanced thinkers ever since people began thinking about leaving for the stars. And who should know this better than Bob Forward, expert in (as he called them) "advanced propulsion concepts."

In the early 1980s Forward had hired himself out as a private consultant to the Air Force Rocket Propulsion Laboratory, getting paid big fees for studying all kinds of exotic methods of propelling rockets through space. Forward's official report, "Alternate Propulsion Energy Sources," was an exhaustive analysis of twenty-six propulsion methods that most people—even including many physicists—had never heard of. First published by the Air Force Space Technology Center at Edwards Air Force Base, it was later reprinted by the trade journal *Commercial Space Report*. Before that, Forward, together with a cadre of other Space Rangers, had compiled long bibliographies of the literature on interstellar rocketry and had published these in several issues of *JBIS*, the *Journal of the British Interplanetary Society*. As was well documented by these lists (which ran to more than seventy pages), chemical rocketry was only *one* way to go, and it was by no means the method of choice; there were countless published articles on nuclear electric and nuclear pulse propulsion schemes, antiproton systems, fusion rockets, interstellar ramjets, and so on. There was even an idea that Bob Forward had come up with himself, back in the early 1960s.

This was for something new under the sun, the so-called rocketless rocket. It would carry no propulsion system whatsoever; the whole thing would be payload.

Clearly, the idea was preposterous: how could you have a spacecraft that lacked a propulsion system? It was preposterous until you thought about the way sailboats had crossed vast oceans with no on-board propulsions systems either. All their motive power came from outside, in the form of wind. Forward's idea was to apply the same principle to interstellar travel, except that instead of wind filling the sails you'd use sunlight.

Solar sailing was an idea that went back to at least 1924, when it had been proposed by the Russian rocket pioneer Fredrich Arturovich Tsander. Later on, a whole assortment of space scientists (including Eric Drexler, who had written a master's thesis on the subject at MIT) had worked out methods of making the solar sail into a paying proposition. Even the sail's greatest advocates, however, had to concede that it would work only within the solar system, the reason being that sunlight got weaker as the square of the distance from the source, making it virtually unusable once you got to the outer planets.

In 1960, however, a light much stronger than sunlight was invented, the laser beam, and almost as soon as he heard about it Forward started thinking about possible applications. Laser beams hardly expanded at all as they traveled: they could penetrate out for thousands of miles and still remain almost as concentrated as they were at the point of origin. If a laser beam were directed against an appropriately sized solar sail, he realized, it would be able to push a spacecraft a good distance out into the solar system— maybe even completely beyond it, over to the next star. All you'd need was a strong enough laser-beam generator at one end and an oversize lightsail at the other.

The energy needs would be immense, certainly, but the sun radiated out a stream of energy more than ample to power an interstellar laser. As for the lightsail on the receiving end, this would be immense by current standards, measured not in feet but in miles; but putting together structures of that size was not an especially daunting task so long as you did it in free space.

There *was* one problem with this setup, however, which was that there seemed to be no way of *stopping* a laser-pushed spacecraft once it got to where it was going; for even if you switched the beam off long before the spacecraft reached the target star, the ship's momentum would still take it past the star and off into the Great Beyond. Likewise, even if by some miracle you managed to get the spacecraft stopped at its destination, there was still no way of using the laser beam to get it moving back toward earth again.

You could avoid these problems, of course, by equipping the spacecraft with conventional chemical rockets, but that was defeating the whole purpose. Besides, it would be cheating. There ought to be a way to arrange things so that you could make the whole voyage by laser beam alone, but even Bob Forward had a hard time imagining how this might be possible. How could a *pusher* beam *pull* the spacecraft back?

Because it was *impossible,* Forward kept thinking about it. Finally, he conjured up a way.

To stop the craft, what you needed was a two-piece lightsail. Part of the sail would separate and move off so as to reflect the laser beam *back* toward the oncoming spacecraft, forcing it to slow down and eventually halt.

"You'd make your sail in two pieces," he reasoned. "One of them's going to be your retro mirror. You'd launch the craft with your laser beam, pushing it out towards the star. As it gets near the star, the sail breaks into two, and the smaller center portion, containing the payload, drops out. The remaining part, which is now a doughnut-shaped ring, is still facing the earth, and the laser beam coming from the earth keeps pushing *that* piece past the star. But in the process, the light also bounces off the doughnut-shaped ring and then onto the back surface of the payload section. This slows down the payload until it finally stops."

And if you made the thing in *three* sections instead of two, you could repeat the sail-separation process, jettisoning the third portion and using it as a mirror to bounce the incoming laser beam back against the payload section. After a while, the payload would start moving, and sooner or later it would get back to earth.

*F*orward took his world-as-engineering-project viewpoint largely from his teacher, Joseph Weber. Weber was something of an original when it came to physics, and in fact he was a physicist only by default, for Weber never even wanted to *be* a physicist, at least not at first. When he got out of the U.S. Naval Academy at Annapolis, he wanted to go for his doctorate in electrical engineering, but as he discovered when he checked into the matter, none of the schools in the area offered Ph.D.s in the subject.

"The first school I went to, to inquire about graduate work," Weber said, "was George Washington University. I was sent to the office of George Gamow, the person who first proposed the three-degree blackbody radiation."

The three-degree blackbody radiation was a fringe of microwave activity at the furthest extremes of the universe, the last remaining trace of the Big Bang. Gamow had proposed the notion back in the 1940s and discussed it with two of his colleagues, Ralph Alpher and Robert Herman. Alpher and Herman went around asking radio astronomers if there was any way to look for this radiation experimentally, and they always got the same answer: No.

"So Gamow asked me," Joe Weber said, "'What can you do?' and I said, 'I'm a microwave engineer; do you have any problem in microwave physics to discuss, something that I could work on as a doctoral project?' And George Gamow said no and he sent me on my way. The man who proposed the three-degree blackbody radiation never thought for a minute that anybody could go out and look for it."

Later, in 1964, the three-degree blackbody radiation was discovered, quite accidentally, by Robert Wilson and Arno Penzias, two Bell Laboratories radio astronomers, and for this accident they received the 1978 Nobel Prize in physics. That was the first time Joe Weber escaped the Big Swedish Award.

"Gamow had *proposed* the three-degree blackbody radiation!" Weber said. "He knew more about it than anyone in the world, *and I could have gone out and found it, in 1949!* But he just went blank and sent me on my way. I'll never forget that interview."

So anyway Weber went off and enrolled in the physics program at Catholic University, got his degree, and then took a teaching job at the University of Maryland. Fortunately, his physics training hadn't destroyed his engineering talent, with which Weber thought he could still discover far-off influences in the universe. He decided to try his luck at detecting gravitational waves.

According to relativity theory, gravitational waves were ripples in space-time caused by violent events out in the Galaxy. For example, if a star went supernova, the explosion would make waves in the surrounding gravitational field just as a stone made ripples on water, and sooner or later some of those gravitational waves would come washing in over the earth. When they reached earth those waves would jostle things around just the slightest little bit, and Joe Weber thought that if he fashioned a sensitive enough experimental apparatus he could use it to detect the incoming ripples as they arrived. He'd put his ear to the cosmic tracks, as it were, and hear the gravitational freight train steaming in from out of the light-years.

Indeed, if it worked, a gravitational wave detector would be a major cinematic achievement, opening up a whole new window on the universe.

"Gravity is a channel of information about the universe," Weber said. "If there are four basic forces in nature, then the gravitational field can tell us 25 percent of all available information about the world. So far, that information has not been available to anybody."

The detector would be a telescope of a sort, one that saw through the gas and dust that blocked great gobs of the universe from optical view. According to theory, at any rate, gravitational radiation penetrated through cosmic dust as if it weren't there.

No one had ever tried to build a gravitational wave antenna before, so Weber had to design one from scratch, and in fact he designed several. The first thing you needed, he supposed, was a body large enough to be affected by the incoming waves. He calculated that a bar of aluminum about the size of an office desk and weighing some three thousand pounds would be fully up to the task. Such a piece of metal would not exactly bob around like a cork, but with the aid of some extremely sensitive motion detectors (piezoelectric transducers), the bar's slightest motions would be measurable, motions much smaller than the diameter of an atomic nucleus. Just about the time Weber was coming up with all this, Bob Forward, who was a grad student in the physics department, came around looking for a doctoral project.

Forward had been a gravity nut even as a kid. He'd once read this science fiction story—it was in *Amazing* or *Astounding* or another one of those mags—where the hero had to find and destroy an invisible alien spaceship. This didn't sound possible: how could you locate something that you couldn't even see? You couldn't, of course, unless there was a way of detecting an object's *mass*, which didn't seem likely. But that's what the hero did. He put together a "mass detector," switched it on, and the thing swung around and pointed to the invisible ship like a compass to magnetic north.

Bob never forgot that one: a mass detector. You couldn't actually *build* one—this was just science fiction, after all—but still, it was a neat trick.

Bob Forward grew up nearby and went to the University of Maryland for his undergraduate work, transferred to UCLA for his master's, and then came back to Maryland for his Ph.D., where he began to take courses from Joe Weber.

Weber's main virtue, in Forward's eyes, was that he regarded everything from an engineering angle. He had this sort of *operational mind-set*, as if he wanted to interact personally with the basic realities of the cosmos. Weber even taught *relativity theory*, of all things, from an engineering viewpoint, talking about masses and forces and fields, instead of the way mathematicians taught it, which was by juggling abstract numbers and symbols on the blackboard.

At any rate, when Weber needed assistants for the gravitational-wave project, Forward was only too happy to volunteer.

"He joined the project as a Hughes Fellow after I got money for it," Joe Weber recalled. "I was enormously impressed by him. He was just a tremendously bright, imaginative guy."

Some of the work Forward did for Weber, though, took more brute strength than imagination. The aluminum bar came straight from the casting forge and was covered with oxidation and slag; Forward was the one who smoothed it out with an electric sander and then polished it the rest of the way by hand.

For the apparatus to work properly, the quietest possible conditions had to be maintained, but unfortunately the gravity-wave detector had to be housed in the very same building where the School of Engineering's brick-smashing tests were conducted on what seemed to be a daily basis. So Weber, Forward, and David Zipoy, a research assistant professor in the department, came in during the wee quiet hours at nights and on weekends to put their ear to the tracks.

Almost at once, according to Forward, the antenna started registering hits, just as if it were bobbing around in a sea of gravitational whitecaps. The question was whether these were true hits or artifacts created by the mechanism itself, instrument noise being registered as false positives. It didn't seem as if there could possibly be this many gravity waves actually coming in because, according to theory, such waves ought to occur only as often as the events that produced them—mainly supernovas, which took place at the sparse rate of only about one or two a year at most—whereas the experimenters were getting apparent hits by the dozen.

Weber and his crew worked for months trying to calibrate the instrumentation, but even after they did so, the antenna seemed to register more events than could ever be accounted for. Forward himself never knew what to think about the embarrassing superfluity of false positives. "Either there's something wrong with the theory of gravity," he said much later, "or whatever's exciting Weber's antennas is not gravity."

Anyway, that was the second time Joe Weber escaped the Big Swedish Award. Weber himself later recalled a World International Conference of Physics where a famous scientist got up, pointed his finger at Weber, said that he'd done nothing right, and predicted that nothing useful would come out of the field for a hundred years. It was some consolation when the Smithsonian Institution asked for Weber's antenna and put it on permanent display at the Museum of Science and Technology.

Inconclusive as it was, the whole episode was a learning experience for Bob Forward.

"I thought about gravity, and the things that made gravity, and the things that responded to gravity, I thought of them all from a mechanistic point of view. I got an intuitive, gut feeling for what these fields looked like, and of how they'd interact. I could actually *visualize* the fields, so that if I had a complicated structure of masses to deal with, it was relatively easy for me to picture the gravity fields around those things."

Later on, Forward would discover how to make gobs of matter do things that most physicists thought were *impossible*. Mass detectors, antigravity generators, time machines—these were just neat tricks in Bob Forward's eyes.

The Hughes Research Laboratories were set up on a cliff overlooking the Pacific, like everything else in Malibu. For the longest time, though, Bob Forward's office was on the side of the building that faced the mountains, and so he'd sit there and work at his desk and occasionally stare off at them.

He'd marvel at those mountains, he thought of how *big* they were . . . so *massive* . . . all that great, motionless mass just sitting there silently and *deforming space-time*. He could visualize their gravitational fields perfectly, just as if he were looking at iron filings tracing out the force lines over the peaks and valleys.

Forward was living in Oxnard at the time and would get to work by driving south along the coast. He made it a point to do his thinking on this drive: he didn't listen to the radio or watch the girls; he just kept up with the traffic and thought about physics. He'd drive along the beachfront for a while, then would reach the point where the mountains came right down almost to the beach, and finally he'd get to where he'd have to drive past Point Mugu.

Essentially, Point Mugu was a big rock. In fact it was colossal: it went up about a hundred feet, a tremendous space-time-deforming glut of matter.

And as Bob Forward drove past that gigantic rock one time he thought to himself: *If I could only shake that rock hard enough it would make lots of gravitational waves.*

If you shook the rock, he thought, then you'd be shaking the rock's gravitational field along with it, and you'd be able to detect the resulting fluctuations with a gravitional antenna much like the one he'd made with Weber and Zipoy, only it could be a much smaller version, almost pocket-sized. With it you'd be able to measure the strength of the rock's gravitational field, and from *that* you'd be able to deduce the object's mass.

It would be another way of putting your ear to the cosmic tracks, another way to sense what was out there without seeing it. You could fly such a device over the lunar surface, for example, and map out its hills and valleys. You could fly it past the asteroids—the ones you wanted to explore, mine, or colonize, L5 style—and take a mass readout. You might even be able to locate an invisible alien spaceship, if you had to. The only thing was, Bob Forward couldn't for the life of him figure out how to get Point Mugu agitating back and forth at the necessary speeds, which were considerable, somewhere up near the speed of light.

But then in a flash everything was clear. *The speed of light* reminded him of *relativity,* so he thought: Why do I have to shake the *rock?* Why can't I shake the *antenna?*

The fact was that jiggling the antenna would have exactly the same consequences as jiggling the rock: in either case the antenna would be moving with respect to the rock's gravitational field, and in either case this relative motion would be measurable.

Not long after that, Forward invented and patented a gravitational-mass sensor, the first of eighteen patents he'd be awarded during his years at Hughes. To test and calibrate the sensor, he and an assistant, Larry Miller, also invented another gravitational gimmick, a gravity-field *generator.* This would *transmit* gravitational waves of a known strength, ones that the mass sensor could be tested against. Forward soon had a contract from NASA to develop the device—technically a gravity gradiometer—to be used during the Apollo program to do lunar terrain mappings. The mass detector was scheduled to fly on *Apollo 18,* but then the whole moon program was scrubbed with *Apollo 17.*

Later still, Forward saw how you could utilize the same principles to play absolutely unbelievable tricks with gravity. If you arranged a few massive objects in the right patterns, for example, then you could actually control the gravitational fields they generated. You might even be able to *flatten out space-time*—which is to say, destroy or cancel out gravity—over a small, well-defined region.

He originally came up with this for a novel, for by this time (it was the late seventies), he had started writing his own science fiction. It was just a hobby at first, but later on he became so successful at it that it became a full-time occupation. His first novel, *Dragon's Egg,* was about what life would be like on a neutron star.

Since a neutron star is an ordinary star after it's collapsed down into a sphere roughly ten miles across, its associated gravitational field is truly whopping, some sixty-seven billion times that of earth. In the book, Forward wanted to have a manned mission approach the star and orbit around it at a height of four hundred kilometers. Because he had decided that he was going to write only hard science fiction, in which everything that happened was consistent with and fully allowable by the known laws of physics, he was faced with the problem of somehow keeping the crew members from being dismembered by the star's gravitational tidal forces, which were stupendous enough to tear them apart.

It was true enough, of course, that orbiting bodies were in free-fall and were therefore technically "weightless," but to be literally accurate this would be true only of an object's center of mass; those portions of the object that were closer to the star would be attracted with a greater force than those farther away, which meant that even a relatively short object, such as a person, would be torn apart like a piece of taffy. Forward calculated, in fact, that an astronaut's lower half would be pulled downward with a force of 202 Gs more than his middle, while the head would be pulled by a force that was 202 Gs less, making for a total differential pull of 404 gravities.

That was the bad news. The good news was that such forces could be counteracted by a series of "tidal compensator masses," as Forward called them, arranged appropriately about the space-craft. These would be positioned in such a way that they'd act in opposition to the neutron star's mass, thereby canceling out the differential attractions, at least over the short distances within the spacecraft's crew compartment.

These compensator masses would have to be substantial objects in their own right, and Forward decided that a convenient way of making them would be to take an asteroid of rather large size—one about two hundred miles wide, perhaps—and compress it down to where it was about thirty feet across. Forward wasn't just guessing at these values, he'd worked out the physics of it all in precise mathematical detail, sometimes to three decimal places. Indeed there were no lengths he wouldn't go to in order to get the fictional details right, all of which can be found, for those interested, in the technical appendices that accompany each of the novels.

Shortly after he finished *Dragon's Egg*, Forward realized that the gravitational tricks he'd come up with would pertain just as well to spacecraft in earth orbit, although of course the forces involved would be on a much smaller scale. Maybe there were some real-world applications here, in space shuttle experiments, for example. Why not write it up for a serious audience?

So Bob took his science-fictional gravitational mass compensator, dressed it up in technical physics lingo (mentioning, however, that the design had been "originally conceived for a science fiction novel"), and sent it off to the premier American journal of physics, *Physical Review*. The journal accepted it for publication and ran it, complete with diagrams, charts, and equations, under the title, "Flattening Spacetime near the Earth." Here Bob showed that "if we place six 100-kg spheres in a ring whose plane is orthogonal to the local vertical and whose center is at the center of mass of the experiment, the gravity attraction of the spheres will produce a counter-tide that can reduce the earth tide accelerations by factors of 100 or more in significant experiment volumes."

By this stage in the game, Forward had the force of gravity essentially tamed. He'd learned how to modify gravity, how to create and destroy it. He'd explained how you can cancel it out, how you can grab hold of that curving space-time and flatten it out like a pancake. He'd even come up with a design for a *gravity catapult*, with which you could heave objects through space on a flexing wave of controlled gravitation. Indeed, there was virtually no gravitational appliance Bob Forward hadn't thought of. The *anti*gravity machines, in retrospect, were almost the least of it.

*A*ntigravity machines were *impossible*, of course, as any number of respectable physicists would go out of their way to tell you. Milton Rothman, for example, claimed in his book *A Physicist's Guide to Skepticism* that "no one will ever build an antigravity machine, simply because no gravitational repulsion exists. Electrical shielding is possible because there are two kinds of electrical charges: positive and negative. By contrast, there is only one kind of mass. Therefore gravity, an interaction between masses, has only one form: a universal attraction. As a result, there is no way of arranging masses so that they do anything but attract each other with the gravitational force. They can't produce a gravitational shield or a repulsive gravitational force."

Certainly past attempts at building antigravity machines did not inspire much confidence that one could be built in the near future, if ever. Back in the 1940s, a man by the name of Roger Babson, who had made a fortune for himself by selling stock market advice, established a program of cash prizes for ways of controlling, harnessing, or reversing the force of gravity. The awards were administered by Babson's Gravity Research Foundation, which also sponsored conferences in which people sat in "gravity chairs," swallowed "gravity pills" to boost their blood circulation, and heard lectures on how to beat gravity at its own game: go upstairs at high tide, for example, when the ocean's mass will give you a slight lift.

Visitors to the foundation's offices would come in to stare in awe at the original bed of their hero Isaac Newton, who had first stated the principle of universal gravitation. (The foundation had acquired the bed, according to one explanation, "presumably because Newton at one time rested on it by the force of gravity.")

Antigravity enthusiasts attached particular significance to certain events as recorded in the Old Testament: for example, the ascension of the prophets, and Jesus himself, into the wild blue yonder. "The incident of Jesus walking on the water should not be ignored," they claimed.

Bob Forward, though, figured out ways of building antigravity machines without restoring to hocus-pocus. In fact, Bob had come up with his first antigravity devices back in the early sixties, when he was still a grad student and working in Joe Weber's lab. He published the designs in the standard journals, one under the title "Antigravity" in *Proceedings of the IRE,* the other as "Guidelines to Antigravity" in the *American Journal of Physics.* You could use techniques based on standard relativistic physics, he said, to create antigravitational forces anytime you wanted to; all you needed was mass, and lots of it, preferably traveling around at high speeds.

"For example, if we accelerate matter with the density of a dwarf star through pipes wide as a football field wound around a torus with kilometer dimensions, then we could counteract the earth's gravitational field for a few milliseconds."

A modest beginning, but there it was. Afterward he'd find lots of simpler ways of making antigravity, some by using only standard Newtonian physics. For example, if you opposed a given mass by another one of equal magnitude, then any object between the two would be suspended as if it were weightless, which indeed it would be.

"Put another planet, with the same mass as earth, above your head," said Forward. "The Newtonian antigravity field of the above-earth will pull you up with the same force as the Newtonian progravity field of the below-earth is pulling you down. The two forces would cancel each other out; over a broad region between the two 'earths,' there would be no gravity. Everyone and everything would be in free-fall."

As for keeping the two planets apart, all you had to do was place them in orbit around each other; that way, the centrifugal force would counteract the force of gravitational attraction between the two bodies, and everything would be perfectly balanced.

This was, at any rate, how Forward managed the trick in his novel *The Flight of Dragonfly,* about a laser-pushed interstellar voyage to Barnard's star. When the crew members got there what should they find but a *double planet*, which is to say, two planets revolving around each other in tight formation. One of them was made of water, the other of rock, and they were so close together that they even shared a common atmosphere. From a distance the two bodies looked like an infinity symbol, and at the point of near-tangency, there was, as Forward described it, "a ring geyser that shot a fountain of foamy water up toward the zero-gravity point between the planets."

Still, as picturesque as it was, placing two planets in orbit around each other was the hard way of making antigravity. Forward's mass compensators were antigravity machines in their own right, for what did they do but cancel out the force of gravity? In fact, it could all be done even more simply just by positioning a lump of superdense matter over the top of whatever it was you wanted to levitate. A lump of such matter would be small enough to maneuver, and you could keep it off the ground by supporting it on pillars, but the net effect would be the same as if a whole planet was overhead. Which is to say that anything that wasn't tied down would rise up and float free.

Bob Forward rather liked what you could do with this.

"Dare we imagine a future where one of the attractions at a Disney park is a Free-fall Pavilion, rising upward on massive swooping buttresses of pure diamond, which support a brilliantly reflecting roof of ultradense matter . . . and under that roof floats a crowd of fun seekers, swimming through the warm air with colorful feathered wings attached to their arms, living out the legend of Icarus for the price of an E coupon?"

TIME SAFARI, INC.
SAFARIS TO ANY YEAR OF THE PAST.
YOU NAME THE ANIMAL.
WE TAKE YOU THERE.
YOU SHOOT IT.

That was the ad placed by a time-travel firm in Ray Bradbury's short story "A Sound of Thunder." It was so easy to go back through time (at least in the story) that big-game hunters routinely crossed millions of years in complete safety and for the price of only ten thousand dollars.

In countless other stories, too, time travel was portrayed as being only slightly more difficult than crossing the room. For example, in *The Time Machine,* by H. G. Wells, a man visits the year A.D. 802,701 and then returns to the nineteenth century to tell his friends the whole story over dinner.

He had constructed the machine himself, he told them, and got to the past merely by pushing a lever. Other than running up against common sense and being subjected to some extremely weird sensations—"I seemed to reel; I felt a nightmare sensation of falling," and so on—the inventor had no particular problem roving through the centuries as if they were miles.

That was the way it was in fiction. In the real world, by contrast, everyone knew that time travel was *impossible*. In fact, it was probably the world's *single most impossible feat*. Or at least that's how it looked at first glance.

The famous "kill-your-grandfather" paradox, for example, seemed to present an absolute barrier to time travel: If time travel were possible then you could go back and kill your own grandfather, with the result that you never would have existed. But if you never existed, then how could you have killed him? And so on.

Indeed it was all pretty mystifying. But as time-travel fans were quick to point out, the fact that you couldn't kill your own grandfather didn't mean that you couldn't go back in time. You might be able to go back only as an *observer*, witnessing past events the way Scrooge saw his earlier self in Dickens's *A Christmas Carol*. Or if you went back and were absolutely intent on killing *someone*, then the victim would have to be somebody other than your grandfather. Or if for some perverse reason you *had* to kill your own grandfather, then you would have to do so only after he'd sired your *father*, because in that case there would be no more paradox.

So the fact was that if you admitted along with Thomas Aquinas that even *God* cannot change the past ("God cannot effect that anything which is past should not have been," he said. "It is more impossible than raising the dead")—even if you *admitted* that—still and all, there were plenty of ways of escaping the ordinary and usual barriers to time travel.

Naturally, considering that it was commonly perceived to be the world's *single most impossible feat*, time travel was not a subject that Bob Forward could afford to pass up. In his view, though, not only was time travel not impossible, it had already been accomplished, at least in a sense.

This was the one-way time travel allowed by relativistic time dilation, a standard item in physics ever since Einstein demonstrated, back in 1905, that as an object approached the speed of light, the time in its frame of reference would slow down or dilate.

A speeding clock, he predicted, would tick more slowly than a stationary one, a prediction that was verified experimentally in 1971 when physicists John Hafele and Richard Keating took two identical clocks, put one aboard a jet aircraft and left the other on the ground, and then compared results. Lo and behold, the moving clock had run more slowly.

Time dilation had become such a cliché, in fact, that when Forward gave his "Far-Out Physics" lecture, he refused to discuss it. "That's so old, I'm not even going to talk about it," he said.

In any event, time dilation was not time travel in any real sense. *Real* time travel meant going *back* in time, traveling back to witness events that had already happened. But as Forward learned when he investigated the matter, relativity theory seemed to permit this too. The first person to make the discovery seems to have been Kurt Gödel, the logician.

Gödel was a colleague of Einstein's and once talked to him about some of the new cosmological solutions to the relativity equations that he, Gödel, was coming up with. In 1949 Gödel published these solutions and elucidated them by stating they showed time travel to be possible. "By making a round trip on a rocket ship in a sufficiently wide curve," he said, "it is possible in these worlds to travel into any region of the past, present, and future, and back again."

Later, even Einstein himself seemed to be wondering just how real time was. "People like us, who believe in physics," he said, "know that the distinction between past, present, and future is only a stubbornly persistent illusion."

Twenty years later, many of the world's broadest-minded scientists were working to destroy what remained of the past-present-future distinction. Hans Moravec, naturally, was in the forefront of these.

Moravec had come up with his first ideas for time travel while he was still in high school. The key to it all, he thought, was space-time, the famous four-dimensional continuum of Einstein. Space and time were not separate entities but formed an integrated web consisting of one time axis and three space axes. But if that were so, then why couldn't you traverse the temporal axis just as you could the spatial ones? What indeed was there to *stop* you? All you needed in order to travel along the time dimension was an extremely large mass, for it was a fundamental principle of Einsteinian physics that mass warped space-time.

On the other hand, when it came to converting these abstract realizations into a functioning time machine, things got pretty murky. There was this idea about using black holes as gateways to the future—that was a big theme in the science fiction of the period—but even Moravec didn't think he'd survive such a passage. Maybe then you could somehow peek through a black hole to view the future. He really didn't know for sure.

Others worked on the problem too, and such progress was made that by the mid-1970s time travel had become more or less an accepted possibility. The most popular method of creating a time machine was by spinning up a body of ultradense matter until the space-time continuum gave way in disgust and allowed people to proceed through it at will.

Such schemes were advanced by Brandon Carter, Frank Tipler, and many others. Of his own scheme Carter said: "The central region has the properties of a time machine. It is possible, starting from any point in the outer regions of the space, to travel to the interior, move backwards in time . . . and then return to the original position."

And of his similar time-travel recipe, Tipler stated: "General relativity suggests that if we construct a sufficiently large rotating cylinder, we create a time machine."

That was what advanced theoretical physicists were saying about time travel well before the end of the century, all of which was quite pleasing to Bob Forward. There was a peculiar consequence of Tipler's scheme that Forward thought especially noteworthy. This was the result that if backward time travel was really possible, then causality itself might have to be abandoned.

"This is a brand-new idea," Forward said after he read Tipler's piece. "I think the most important thing about it is that if this concept holds up, causality is dead!"

Indeed, matters had now reached the stage where many physicists thought it was folly to claim that *anything* was impossible.

Who could tell, in advance, what kind of weird surprises unaided nature, or nature aided and abetted by human cleverness, might turn up?

The worst thing about time travel, after all, might not be the logical paradoxes involved or the contra-causality or the overthrow of relativity or anything of the sort. The very worst thing might be much more prosaic. It might be that time travel would require too much energy for it ever to take place. Vast amounts of energy would be required: you had to get lots of superdense matter flowing at near-light speeds, and that would be expensive by any measure.

There was even a story written about this, called "The Man from When," by Dannie Plachta. It was about time travel. One day when everything was perking along in its usual and ordinary fashion, a man suddenly materializes on the scene from the indefinite future.

This is how it would have to happen, of course, someone just showing up out of nowhere.

But the time traveler brings with him a message, namely that the energy required to send him back was so great that it *completely destroyed the earth of his own day.*

The time traveler then informs his listeners that he has traveled back a total of eighteen minutes.

Surprise!

All that stuff about spinning up gobs of superdense matter and so on, that was the hard way of traveling through time. As Hans Moravec saw it, there was another way of managing the same trick, based on an entirely different principle. If we can't go back for some reason, maybe we can *bring the past to us.*

It ought to be possible, Moravec thought, to resurrect past history, or at least to resurrect some of the important historical figures—Isaac Newton, for example. In fact, Moravec had already figured out how to do this by the time he'd written *Mind Children.* He'd worked out a whole scenario whereby a powerful enough supercomputer would be able to resurrect long-dead minds from the information that still survived. You might be able to get Isaac Newton back from an edition of *Principia Mathematica,* plus what flimsy disturbances might remain in the air from the words Newton had actually spoken during his lifetime.

This was a bit of a stretch, admittedly, but nothing absolutely impossible, at least not in Hans Moravec's view. After all, plenty of archaeologists had made a living by reconstructing entire cultures from pottery shards, scraps of ancient documents, X-ray scans of mummified remains, and so on, so why shouldn't the superintelligences of the future be able to go far beyond this, to the point "where long-dead people can be reconstructed in near-perfect detail at any stage of their life," as Moravec put it? It would be another fun project, along with these robots.

"It might be fun to resurrect all the past inhabitants of earth this way and to give them an opportunity to share with us in the (ephemeral) immortality of transplanted minds."

And as it turned out, just as he was coming up with this so was another advanced thinker, Frank Tipler of Tulane University, the very same physicist who had, with John Barrow, made all those predictions about life occupying every region of all possible universes. Like Moravec, Tipler presupposed a really great supercomputer that would allow you to simulate virtually anything. Once you had such a thing, you could bring back the past without any trouble, by simulating the mind as well as the body of the dead person.

"Simply begin the simulation with the brain memory of the dead person as it was at the instant of death (or, say, ten years before or twenty minutes before) implanted in the simulated body of the dead person, the body being as it was at age twenty (or any other age). This body and memory collection could be set in any simulated background environment: a simulated world indistinguishable from the long-extinct society and physical universe of the revived dead person; or even a world that never existed, but one as close as logically possible to the ideal *fantasy* world of the long-dead person. Furthermore, all possible combinations of resurrected dead can be placed in the same simulation and allowed to interact."

Much like the computer virus that Moravec was writing a science fiction novel about (in his collaboration with Fred Pohl), these simulated people's lives could be run back and forth like a movie film. Later, Tipler considered the objections that had been made in the past to religious views of resurrection. David Hume, for example, said that resurrection was impossible, or at least highly unlikely, because of the fact that there were *too many dead people,* and, what was even worse, many of them were *not worth resurrecting.*

"A great proportion of the human race has hardly any intellectual qualities," said Hume. "Yet all these must be immortal. A Porter who gets drunk by ten o'clock with gin must be immortal; the trash of every age must be preserved, and new Universes must be created to contain such infinite numbers."

David Hume could not accept this, seeing it all as "a most unreasonable fancy," but Frank Tipler had no trouble with it. "The ever-growing numbers of people whom Hume regarded as trash nevertheless could be preserved forever in our single finite (classical) universe if computer capacity is created fast enough." Eventually, Tipler was sure, "the computer capacity will be there to preserve even drunken porters."

And for those who were *still* skeptical that these oceans of past people could be resurrected by means of computer simulations, Tipler had a Final Knockdown Argument. This proof was so strong that it covered not only *past* humans but *all possible humans.*

"Merely simulate all possible life-forms that could be coded by DNA (for technical reasons, the number is finite), and all possible humans necessarily will be included. Such a brute force method is not very elegant; I discuss it only to demonstrate that resurrection is unquestionably physically possible."

*I*n 1987, Bob Forward took early retirement from the Hughes Company. He had worked there for thirty-one years, had been awarded eighteen patents on behalf of the company, had published countless articles and a few books, had delivered countless talks, and was now hard at work at science fiction. At the time he retired, Hughes had started publishing Forward's own personal scientific magazine, called *Mirror Matter Newsletter,* which was devoted to his far-out antimatter ideas. (Forward calls antimatter "mirror matter.")

So much imagination! So many weird schemes! So many inside

jokes! Like the headline to one *Mirror Matter* story: "?EMAN A NI S'TAHW"

The story: "In electrical engineering, the unit of conductance is the MHO, while the unit of electrical resistance is the OHM, and conductance is the inverse of resistance. Thus, it is obvious that this inverted MATTER stuff we are trying to find a name for should be called RETTAM." (This was an inside joke.)

You had to wonder why his mind didn't crack from the strain of it all. Eventually, Bob Forward decided that enough was enough. He had more ideas than he could possibly put to good use at the Hughes Research Laboratories, so he left to become a full-time writer.

He divided his time between two houses, one at the edge of a cliff in Idyllwild, California, up at about six thousand feet, the other an old farmhouse in Scotland, which overlooked the North Sea.

He started out wanting to conquer the impossibles ("I always had it in the back of my mind some way to go faster than the speed of light, some way to go backwards in time"), and by the time he left Hughes he'd done it all, at least conceptually.

Which led to the question of what the human race would ever do once its finest hubristic thinkers—the Bob Forwards, the Hans Moravecs, the Eric Drexlers, the Dave Criswells, the Frank Tiplers, and the rest—had solved all the world's problems. What would people do on the Monday morning *after* the Barrow-and-Tipler Omega Point had been reached, *after* we'd "gained control of *all* matter and forces not only in a single universe, but in all universes whose existence is logically possible"?

What would be left to do *then*?

9

Laissez le Bon Temps Rouler

*I*n 1987 Keith Henson founded the Far Edge Committee. The group's sole purpose was to begin planning for the Far Edge Party, an enormous gathering of downloaded multiple selves that was to be held in the far-off future and at the other side of the Milky Way.

Henson came up with the idea after realizing there was no way that he personally was going to make a grand tour of the Galaxy if there was only one copy of Keith Henson alive, even if that copy was supposed to live forever. *He* might live forever, but the Galaxy sure wouldn't.

"There are 100 to 200 billion stars in our galaxy alone," Henson said, "and even with nanotechnology to help, it will take a year or two per star system, not counting travel time between stars. Visiting every interesting object in serial is literally impossible, since the interesting places won't last long enough. I don't want to take such a long time looking over this one small flock of stars that most of them burn out."

Plainly there was a problem here: how could a single person see all there was to see if part of your destination went up in smoke while you were still in transit? You couldn't, of course, not if you were just *one* person. But if you were *many* people—a bunch of parallel selves—well, then, that would be a different story. In that case your different selves could visit the Galaxy's major hot spots *simultaneously*, before any of those great cosmic tourist attractions had a chance to evaporate. That way, *all* of you could see *everything*.

"We won't have to make hard choices about which way to go," Keith Henson said. "We take all roads."

Later, after your multiple personalities had collectively experienced all there was to experience, you could get them back together again to communicate and share memories. Each of you would behold what every other self had seen and done, so it would be as if a single person had in fact done it all.

Those were Keith Henson's travel plans for the immediate future—for the next million years or so—but as extravagant as it all was, none of it was in the least impossible, not in the age of downloading, when every smart person had made and stored away dozens of duplicate selves already. All you had to do was make a number of *extra* copies of yourself, place them in assembler-made starships, and send them winging to all corners of creation. Those supplemental downloaded minds would stop at every new star system for as long as they cared to and, in their spare time, would make even *more* copies of themselves, as well as their spaceships, all of which would then rocket away to the next nearest solar system, where the whole process would repeat itself. Making new spacecraft wouldn't be any problem either, not in the era of advanced nanotechnological excess, when unending quantities of anything could be had for virtually nothing

"So we will sweep across the Galaxy," Keith Henson said, "and converge for a giant party, scientific meeting, and for those who want it, a memory merge, so they can have seen all the wonders of the Galaxy."

Naturally, there were a few problems still to be worked out. For one thing, a clean sweep of the Milky Way might take quite some time—a full two hundred thousand years even at half the speed of light. How could all the multiple selves be expected to show up simultaneously after all those years? And where would they meet?

"How do we get them all together at the same place?" Henson wondered. "How do we get back together at a place we can't even *see* from here? Should we give one party per galaxy? Or one on the far side of the Virgo cluster? How many centuries should we party? How much bean dip will we need?"

*F*or a long time even the most gung-ho optimist had to admit that the great hubristic adventure would have to come to an end when heat-death time rolled around. No matter where we might have gotten to in the interim, no matter what miracles we might have accomplished out there in the universe at large, the heat death ultimately would put an end to humankind, progress, and every-thing else, once and for all. As far off as it was—untold billions of years in the distance—there seemed to be no way of escaping the final fade-out.

The idea that the universe would eventually run out of steam was first advanced in1854 by the German physicist Herman von Helmholtz, who claimed in an article titled "On the Interaction of the Natural Forces" that sooner or later all available heat would radiate away out into the cosmos, leaving major downtown areas cold and dead. This dim picture had emerged, in turn, from a close reading of the second law of thermodynamics, sometimes called the law of dissipation, which had just recently been announced by another German physicist, Rudolf Clausius. The law stated, in essence, that whenever heat was made to do work, some of it inevitably got away, dispersing off into the environment, and good-bye. Put Clausius and Helmholtz together and it was a death sentence for Mother Nature.

News that the cosmos would one day expire came as a shock to those who kept abreast of the progress of science. This was, after all, the century in which Darwin had announced the theory of natural selection, according to which animals got fitter and stronger, species got better adapted to their environment, and so on, and now it suddenly emerged that in the end none of it would make any difference.

Later, it's true, some thinkers of a cynical turn of mind managed to find some redeeming social value in what was otherwise a rather dark prospect. Bertrand Russell, for example, allowed that "although it is a gloomy view to suppose that life will die out, sometimes when I contemplate the things that people do with their lives I think it is almost a consolation."

That was putting the best face on things, which was, at the time, about the most anyone could have done. This uneasy situation lasted for some years, until the 1970s, at which time even *worse* news arrived from the world of theoretical physics: the proton would decay, and this would happen far in advance of the heat death. Atomic matter would keel over and die long before the heat death even got its chance. Even the particle physicists, the very people who had unearthed this mischief, were quite unhappy with the new outlook.

Howard Georgi, the Harvard physicist, remembered the night he was working out the details of the first Grand Unified Theory, the one he'd collaborated on with Sheldon Glashow, and saw that the particular model he was now studying looked quite promising.

"The model worked," he said. "Everything fit neatly into it. I was very excited, and I sat down and had a glass of scotch and thought about it for a while."

And as he drank his scotch and thought, Georgi realized that this model, which was clearly preferable from every other angle, came packaged with a distinctly unfortunate consequence.

"I realized this made the proton, the basic building block of the atom, unstable. At that point I became very depressed and went to bed."

And who could not sympathize? Proton decay, after all, was the specter of matter disintegrating and turning to mush. Physical objects would collapse in a heap, just as if they were made of sand. Nothing built out of atoms would survive, least of all human beings. The only bright side to the picture was that none of this would happen for a *long* time, some 10^{31} years *at least*.

"It wasn't shattering," said Sheldon Glashow. "I mean, we know the sun will burn out in a few billion years. This is *known*. It's a *fact*. Spaceship Earth and all that—poof! That matter falls apart a long, long time afterward is scarcely an upsetting idea. It's bad enough as it is."

But then Freeman Dyson turned to the problem.

As he searched the physics literature on the long-term future of the universe, Dyson noticed that the available papers on the subject shared a certain strange peculiarity. "The striking thing about these papers," Dyson recalled afterward, "is that they are written in an apologetic or jocular style, as if the authors were begging us not to take them seriously."

It was not a proper use of your time, apparently, to imagine what might or might not happen to the universe some billions of years down the road—a prejudice that was rather surprising in view of the fact that many physicists nonetheless lavished huge amounts of recycled paper, time, and attention on what had happened billions of years in the *past*. Whatever the explanation for the disparity, Dyson noticed that even those who *were* brave enough to think about The End of the World nevertheless suffered from a profound philosophical blind spot in their approach to the subject. All of them seemed to imagine that you could project how it would all end simply by taking known physical laws, applying them to the current state of the universe, and then extrapolating out as far as you could.

That was not, on the face of it, an unreasonable way to proceed: What *else* could you do other than follow nature's laws to wherever they might go? Still, that way of thinking left something crucial out of the picture, something that was piercingly obvious to the man who had dreamt up the "Dyson sphere," who had said there were *too many stars in the Galaxy,* and who had imagined arranging collisions between excess stars in order to reduce their numbers to acceptable levels. What that whole approach omitted was the role that conscious intelligence might play in the way things turned out. Dyson now stated what was, in hindsight, seen at once to be an unmistakable truth: "It is impossible to calculate in detail the long-range future of the universe without including *the effects of life and intelligence.*"

This was a fundamentally new idea. Left to its own devices, the universe might very well come to an end, but put life and intelligence into the picture and . . . well, who could say what the results might be? At the very least, intelligence expanded the range of possible alternatives. Through adroit use of resources, possibly the heat death could be postponed. Maybe it could be eliminated altogether. Maybe even the proton-decay problem could be gotten around somehow. What was for sure was that the undertakings of conscious, intelligent beings would have to make *some* difference to the way things turned out.

Much would depend, Dyson knew, on whether the universe was open or closed. Conventional wisdom had it that a closed universe, one that was limited in space and time, would collapse in on itself, with humanity and everything else perishing in a last gasp of fire and flame. Dyson hated the very idea of a closed universe: "It gives me a feeling of claustrophobia," he said. He therefore considered whether there were ways of actually *opening up* a closed universe, unlocking it, prying its jaws apart like a great cosmic clamshell.

"Is it conceivable," he wondered, "that by intelligent intervention—converting matter into radiation and causing energy to flow purposefully on a cosmic scale—we could break open a closed universe and change the topology of space-time so that only a part of it would collapse and another part of it would expand forever?"

That was a tall order, this business of *causing energy to flow purposefully on a cosmic scale, changing the topology of space-time*, but so what? What if it could be done? It would be a "technological fix," Dyson thought, one that would "burst open" the cosmos like a gigantic seed pod so that life and intelligence could blossom out once again. At all events you wouldn't have to stand by helplessly as the universe collapsed to nothingness in front of you—the way those poor astronauts did in Poul Anderson's novel *Tau Zero*, for example. They'd motored off to the edge of creation in their warp-drive spaceship and then stood back and watched in horror as everything else went up in smoke before their very eyes. All that was *so unnecessary!* Why tolerate such a fate when you clearly had a chance to *do* something about it?

Cosmic energy flows, of course, were needed only for a *closed* universe; an open universe, by contrast, presented an entirely different set of problems. For one thing, you wouldn't have to pry it apart; if anything you'd try to keep it from expanding beyond a certain point (if that were possible), in order to retain the heat. The challenge in an expanding universe would be accommodating oneself to gradually falling temperatures. Maybe the best thing here, Dyson thought, would be to make changes to the human body itself. If a warm-blooded species wouldn't be comfortable in a cold-blooded universe, then the logical thing to do would be to mutate that species into something more fitting. The only question was, what?

Dyson was mulling all this over at about the same time that the idea of downloading was in the air—it was the late seventies—and as Dick Fredericksen, Bob Truax, Hans Moravec, and others were also doing, concurrently, Dyson now wondered whether a human being's mental structures couldn't be implemented in something other than brain tissue. He wasn't thinking in terms of computers, necessarily, but of more celestial objects—things like interstellar dust clouds, for example. Fred Hoyle, the astronomer, had written a novel called *The Black Cloud,* in which cosmic dust managed to evolve life and intelligence. Maybe we could evolve in that direction.

"We cannot imagine in detail how such a cloud could maintain the state of dynamic equilibrium that we call life," Dyson admitted. "But we also could not have imagined the architecture of a living cell of protoplasm if we had never seen one."

At least the idea wasn't impossible. Anyway, the operative point was that if human beings could refashion themselves as dust clouds or something similar then they'd have a fighting chance of surviving the great metropolitan ice age. The pace of cosmic life, such as it was, would be slowed down a bit by the cold, but did that really matter? The fact was, no one would *notice* the change, this as a result of a "biological scaling hypothesis" that Dyson now formulated.

Dyson's scaling hypothesis stated, in essence, that even if your physical functions were restrained by declining temperatures, your conscious experience wouldn't necessarily be affected. There was a distinction between *physical time,* which was measured in minutes and seconds, and what Dyson called *subjective time,* which was measured in "moments of consciousness." This difference between the objective and the subjective would allow your conscious experience to last indefinitely, even though your physical life span might still be finite. Even if external events happened at progressively slower rates, your moments of consciousness would *seem* to occur just as often as they had before. So as far as you yourself were concerned, you'd be immortal.

And when the universe got *very* cold, there was still one last trick to play. What did *animals* do, after all, when it got too cold for normal life? The answer was, they hibernated. Well, intelligent dust clouds could do that too: they could *live in spurts*.

The advantage of hibernation was that you'd hardly notice the downtime, because of course you wouldn't be awake to experience it. From what you could tell out there in interstellar cloud land, things would still be happening, your vast cosmic thoughts still flowing.

"This example shows," said Dyson, "that it is possible for life with the strategy of hibernation to achieve simultaneously its two main objectives. First, subjective time is infinite; although the biological clocks are slowing down and running intermittently as the universe expands, subjective time goes on forever. Second, the total energy required for indefinite survival is finite."

In fact, not only was it *finite,* the energy required was not even very great—"about as much energy as the sun radiates in eight hours," Dyson said.

That was the way to beat not only the heat death, but also proton decay. So long as you could organize yourself in a form that required no atoms—as an electron plasma, say—proton decay would be of no special consequence.

Personally, Dyson didn't expect the proton to fall apart. "It is difficult to see why a proton should not decay rapidly if it can decay at all," he said. As it was, the proton was thought to have an average life span of 10^{31} years.

"So far as we can imagine into the future," Dyson concluded, "things continue to happen."

With Dyson having set such a splendid example, other physicists now announced their own proton-decay work-arounds, heat-death escapes, and generalized universe-saving strategies. In 1982, for example, Steven Frautschi, a Caltech physicist, figured out how in an expanding cosmos "life might attempt to maintain itself indefinitely, and even play a major role in shaping the universe."

As the cosmic juice ran out, he said, a "sufficiently resourceful intelligence" (he didn't speak of "people" in this context) would go around the universe rounding up black holes, towing them into position, and living off the energy they emitted via Hawking radiation. Frautschi's argument, advanced in the pages of *Science*, the mainstream American scientific journal of record, came fitted out with all the standard physics paraphernalia, including formulas for the energy costs of towing the black holes from place to place, the preferred radius of the final black hole collection (an assemblage that he referred to as *the empire*), the anticipated radiation release rates of the black hole empire, and so on.

"It stands as a challenge to the future," Frautschi conceded, "to find dematerialized modes of organization (based on dust clouds or an e^+ e^- plasma?) capable of self-replication. If radiant energy production continues without limit, there remains hope that life capable of using it forever can be created."

Two years later, in 1984, Bob Forward provided an alternative scenario, one that didn't even require hauling black holes all over creation. Forward thought that you could get energy out of the void itself: you could put the very interstellar vacuum to work.

He published the suggestion in *Physical Review* under the title "Extracting Electrical Energy from the Vacuum by Cohesion of Charged Foliated Conductors." It was almost like getting something from nothing.

"If you put a piece of metal out in the vacuum," he explained, "some radio waves are going to be made out of 'nothing' and they're going to bounce off the metal and give it a push. Now there will be waves coming onto the other side too, so there's no overall net resultant, but if you put two sheets of metal facing each other, then they'll be pushed together by the Casimir force."

The Casimir force was a little-known but nevertheless quite real phenomenon that worked across short distances to attract objects to each other; it was an effect similar to surface tension. Objects in motion meant *kinetic energy,* and kinetic energy meant *useful work,* so from two pieces of metal hanging out there in the vacuum there was suddenly useful work being done. "That's not perpetual motion," Bob Forward said, wishing he could polish off *that* impossible feat. "But it's close."

Still later, in 1988, Hans Moravec came out with *his* game plan. In a way, it was the most obvious gimmick of all. You simply extracted the universe's remaining energy while it was still there to be had, and then you deposited it in a *battery* of some kind, a cosmic storage bank. That way, no matter what happened to the rest of the universe (and who really cared?), you'd still have something left in the bank, a power source that could be made to last forever, so long as all you wanted to do with it was *think*.

"The idea is to use about half the energy in the battery to do T amount of thinking," Moravec said, "then to wait until the universe is cold enough to permit half the *remaining* energy to support another T, and so on indefinitely. In this way a fixed amount of energy could power an unlimited stretch of thought."

Lord, what fools these mortals be! Having their heads cut off, turning themselves into computers, beaming themselves across the Galaxy as if they were just so many loose photons.

It was a little touching, of course, to see these people hopelessly striving after the impossible, reaching out in their different quixotic ways against the final hostility of the universe. It was poignant, the hopes they had and the dreams they dreamt—all of which were unfortunately doomed to extinction.

That was a skeptic's view of the situation, and quite understandable it was, too, for how many times in the past had people's boldest hopes and dreams been turned to dust by the cruel forces of nature? Indeed, the folly of tempting God and the fates had always been a primary theme in world literature: Adam and Eve, Oedipus, Prometheus, Faust, Ahab—all of them went up against the gods and then got damned to hell for their arrogance. Tragic flaw that it was, hubris was a common literary theme precisely because it seemed to be such an inescapable part of the human condition.

But none of our extremely advanced thinkers ever saw themselves as tragic heroes. And why should they? In their own eyes, at least, all they were doing was using science in the ordinary way, to gain control over nature and improve the lot of humanity, just as their predecessors had done. Extreme and exotic though they were, not one of their hubristic schemes violated known physical laws. Just the reverse: these scientists started with the known laws of nature, which they then applied to their own purposes.

Wasn't this what scientists had *always* done? Wasn't this what they were *supposed* to do?

Ah, but the skeptic might see an element of denial in operation here. There was something *unseemly* about the way these people wanted to escape from the earth, leave the solar system, and end up over at the far edge of the Galaxy. Didn't this stuff betray a certain . . . *pathology of the intellect?*

Not according to *them* it didn't. The fact was that people had *always* wanted to escape, whether from the cave or from wherever else they'd gotten to. The nomadic hunter-gatherers, the New World pilgrims, the great explorers, the pioneers of the American West—what had they done but leave the scene for greener pastures elsewhere? Some of them had been motivated by escape from oppression, but all of them must have felt the pull of new horizons, the call of the frontier, the lure of pushing outward, of going beyond the edge. Whatever it was, the tendency to *go beyond* seemed to be so deeply rooted in the human genes that it was hard to find anything wrong with it. It, too, was the common lot of mankind, part and parcel of the human condition.

Ah, but the skeptic might now observe that there was something dark and desperate about the new picture we had before us, where people were proposing schemes for . . . *becoming supermen,* making themselves into *transhumans,* and God only knew what other insane blasphemies. No longer were these people merely crawling out of the cave or going to where the grass was greener: they were tampering with human nature itself.

But there was nothing new about that, either.

The term *superman* went back to the second century A.D. when Lucian of Samosata, the Greek satirist, spoke of a *hyperanthropos,* a more-than-human, or "hyper-man." As for the underlying supposition involved—the notion that human nature was not good enough, that man had to be *improved upon*—this point went even further back, to the very dawn of ethics. Hadn't man always been taught that he had to "grow," intellectually, spiritually, and morally? Hadn't he been criticized for being "materialistic"? Hadn't he been instructed to transcend his animal instincts, primal urges, and dirty bodily lusts? Hadn't he been told, in short, to make fundamental enhancements to his vile, vulgar self? But what were all these exhortations other than attempts to get man to surpass what he already was, to eclipse himself, to go beyond, to become more?

And as for the favored end-state, the condition toward which man was supposed to be pointing himself, hadn't it always embraced those very same glories that our crazed scientists now held out in promise—blessed states like immortality, pure spirituality, and perfect understanding? Weren't these in fact some of the oldest and highest dreams of mankind? And were our manic scientists—the cryonicists, the nanotechnologists, the computer downloaders—were they guilty of anything more than the sin of taking exactly those desires seriously?

Man had *always* been aiming at some type of transhuman condition, whether on earth or in heaven. But if the desire to become more than human was as old as the hills, it was nevertheless true that a new element *had* been brought into the picture by the cryonicists, by Eric Drexler, by Hans Moravec, and the rest: what they'd given us (or soon would) was the ability to *satisfy* those desires. No more would all this stuff about "transcending ourselves" be hollow rhetoric. No more would it be mere homiletics, empty catchphrases, and recycled metaphors. Now all of it could be taken seriously, it could be taken *literally,* for if these over-the-edge thinkers were right, then we could achieve those consecrated states quite soon enough—in *just fifty years,* according to Moravec. That was when we could expect to become pure spirits, the superhuman intellects that would outlast the heat death.

The irony of it all was that it wouldn't be *religion* that would give us this ability; we wouldn't be getting it from the supernatural, or from voices out of the crypt. Rather, we'd gain that ability simply through the normal and ordinary progress of science. Just plain science would give us the chance to surpass our old selves, leaving behind our crass materialism and all the rest of that excess baggage.

So, in all reason, you couldn't hold it against our hubristic thinkers—inventors of everything from downloaded minds to Great Mambo Chicken—you couldn't hold *them* responsible for wanting to be more than they were. They weren't Martians, after all; they too were *only human,* and so they wanted the same things that everyone else did.

What they *would* be responsible for (if they had their way), was bringing science to the point where it would finally have caught up with what it was that human beings had always craved, which is to say, immortality and transcendence. They'd take science to the point where the human species could step right up to the edge, look across the dividing line separating the Human from the Transhuman . . . and then cross over.

Brave hopes, certainly, but truly advanced thinkers never were afraid of the penalties attached to hubris. Hans Moravec, in particular, seemed undaunted—and in fact almost challenged—by the antihubristic outpourings of the ancients.

"One day we'll see if the Greeks were right," he said.

And now with the heat death left behind, proton decay outwitted, energy supplies to last forever, and the problem of hubris solved, there was really nothing left to be done. The Barrow and Tipler prophecy was on its way to being completely fulfilled: "Life will have gained control of *all* matter and forces not only in a single universe, but in all universes whose existence is logically possible; life will have spread into *all* spatial regions in all universes which could logically exist, and will have stored an infinite amount of information, including *all* bits of knowledge which it is logically possible to know."

There was nothing to do now but let the good times roll.

For a while Keith Henson thought of establishing a Last Proton Club, the members of which would gather together for the purpose of watching the last proton decay (if they could find it). "This *does* present a few problems," he said. "Like what material form do you take to watch the last piece of nuclear matter decay? But we have lots of time to solve those problems!"

Indeed, if there was anything Keith Henson believed, it was that *all problems could be solved;* so he and his friends turned their attention, as time and energy permitted, to the more pressing business of planning for the Far Edge Party.

"Everybody I know is going to be there," Henson said. "I must know about a thousand people who are all planning to come."

By party time, of course, that initial thousand would have burgeoned into countless trillions of people—what with all the downloading and splitting and copying and so on—making for logistics problems of considerable dimensions.

"If the party got big enough," he said, "the bean dip alone would form a black hole." (Among party organizers, this was known as *The Bean Dip Catastrophe*. Another nightmare was that every third person would turn out to be Keith Henson.)

"Where do you park fifty billion starships?" he continued. "Where are you going to find a big enough party hotel?"

You had to think big when you were dealing with revelry on this scale. It was no place for timidity—not that Keith Henson was much given to that.

"I expect to convert a whole galaxy into beer cans," he said.

Epilogue

In the fall of 1988 Bob Truax sold his homemade launch vehicle, the Volksrocket X-3, also known as Project Private Enterprise, to the United States Navy. The purchase price was $750,000.

For some unfathomable reason that even he could not figure out, the government was now hotly interested in precisely the kind of sea-launched and recoverable rocket that Truax had proposed to much laughter back in the 1960s. In fact, the navy had come up with a new concept called SEALAR, standing for Sea Launch and Recovery, the object of which was to place a ten-thousand-pound payload into low-earth orbit using a two-stage rocket that would be launched from the ocean, landed back in the water some four hundred miles downrange, and then reused.

It was a Truax dream come true, and when the navy let it be known that it needed a specimen rocket for testing purposes, Bob Truax proposed his X-3. In a very few months the navy went ahead and bought it lock, stock, and barrel. They got the rocket, the transporter, the ground support equipment, the computerized control panel, and all other supporting gadgetry and appurtenances. This was the very same rocket Truax had put together in his garage, using the surplus engine parts he'd rescued from the scrap heap—the very same Atlas vernier engines that he'd bought for $25 apiece at a junkyard after the government had spent millions developing them—and now the government was buying it all *back* for $750,000.

Truax didn't think this was a lot of money, not in view of what it represented. "It's for twelve years of work," he said. "It's pay after the fact at some yearly rate or other." Indeed, it worked out to $62,500 per annum, before taxes, expenses, and so on, not really an extraordinary sum.

The private astronaut project, naturally, went on the back burner. "The navy's not interested in it," Truax said, "and neither is anyone else at the moment. No one who has any money, anyway."

Evel Knievel, meanwhile, had long since retired from canyon-jumping and all other forms of low-altitude rocketry and now spent his time traveling, playing golf, trying to sell his original art works—oil paintings, as he'd become an amateur painter—and giving his son Robbie, who was also a death-defying motorcyclist, career advice. One idea was for Robbie to try and jump Snake River Canyon, and every so often Knievel would get in touch with Truax and ask him about it, but Truax wasn't interested.

"I don't think I will," Truax said. "They always want to do it on a cheapo basis—you know, nickel and dime you to death—and that runs the risk that I'd kill the poor kid. I don't want to participate in anything that's going to be *real* hazardous."

By mid-1989 Carolyn Henson, who had changed back to her maiden name, Carolyn Meinel, moved to Albuquerque, where she became a consultant to the University of Arizona Center for the Utilization of Local Planetary Resources, a research center that had been funded by NASA to develop space-mining technologies. Carolyn had become convinced that conventional chemical rockets had become far too expensive to launch a human migration into space, and was sure that everything now hinged on the invention of radical new launch technologies. She wasn't expecting to go up into orbit anytime soon, but she always closed her letters with "Reach for the stars!"

Keith Henson, in San Jose, made a living doing software programming. He was a partner in a firm called the Grasshopper

Group, which had developed a program, MacNews, which they marketed as the first window system for Macintosh Unix.

That's only what he did for a living. For entertainment Keith Henson hassled government bureaucrats, especially those who were involved in cryonics matters. This went back to the way he felt during the Dora Kent crisis: indeed, he felt that his only chance for personal immortality was being threatened. "Some of us were frantic, absolutely frantic, that state government was going to legislate cryonics out of existence," he said. His phone bill had gone up a thousand dollars a month during the worst of it.

Gerard O'Neill, the man who had started the space-colony movement in 1974, went on to make a small fortune with a company he founded, Geostar, which marketed a satellite-based navigation and communications system that he invented. In 1983, when he formed the company, he sold stock to friends and relatives for a penny a share. By the end of that same year shares were trading at thirty-two dollars each.

OTRAG, the German rocket firm that had operated out of Africa, finally closed up shop and placed its remaining stores of rocket parts behind lock and key in a Munich warehouse.

Jim Bennett, who had watched O'Neill give a talk at the University of Michigan and then worked for a succession of private rocket companies, became president of American Rocket Company after its founder, George Koopman, was killed in an auto accident in July 1989 while he was on his way to a static test. Carolyn Meinel went out and attended the funeral. The AMROC launch vehicle—the guaranteed nonexplodable rocket—was renamed the *Koopman Express* in his honor. Carolyn went out to see the first flight test too, at Vandenberg Air Force Base, in California.

On October 5, 1989, the *Koopman Express* burned up (but did not explode) on the launchpad.

*T*wo years after the Dora Kent case, none of the people who participated in her suspension had been indicted for murder, and it was getting to be increasingly doubtful that any of them ever would be. The case had gone to the Riverside County Grand Jury, which asked the defendants to testify; but three of those involved, Hugh Hixon, R. Michael Perry, and Scott Green, all of whom were present in the operating room the night Dora Kent's head was removed, invoked their right against self-incrimination and declined to do so. The prosecutors granted them limited immunity and requested that the judge order them to testify, but the judge refused to do this on the grounds that state law required the defendants to be given *total* immunity. The prosecutors, however, had insuperable problems with this.

"Having no idea as to who may have injected barbiturates into Dora Kent or how it may have happened," said Deputy District Attorney Curtis Hinman, "we cannot grant total immunity to a potential murderer."

Then again there was the chance that if the defendants *were* given total immunity, one of them might falsely confess just in order to save the others from being charged. "Someone might be willing to be a martyr," Hinman said, "because Alcor is in some respects like a religion. They offer life after death."

The Alcorians were still in legal limbo, but meanwhile the moon-lighting deputy coroners ("Riverside's Fun Couple") were cleared of wrongdoing. Brad and Didi Birdsall, coroner Ray Carrillo finally decided, "did not break any laws that we know of."

In 1989, the Alcor Life Extension Foundation received a legacy from one of its patients, Dick Jones, who had died in December 1988. Jones, an Emmy Award–winning television producer, left an estate valued at some 10 million dollars, and half of it would go to Alcor.

At Alcor, though, it was business as usual. Mike Darwin, Keith Henson, Brenda Peters, and others were still going around to conferences and conventions, especially science fiction gatherings, trying to round up new clients. The Alcorians were a big hit at "Westercon 42," the forty-second annual West Coast Science Fantasy Conference, which was held in San Francisco over the Fourth of July weekend, 1989. They set up an information booth displaying a new sign that Brenda Peters had come up with:

Alcor
"Deathbusters!"
("Who ya gonna call?")

Back in Oak Park, Michigan, Bob Ettinger was still publishing his monthly magazine, *The Immortalist,* and coming up with some great new quips, such as "Frozen people are *relatively* dead; Albert Einstein is *absolutely* dead."

*I*n 1989 Eric Drexler and his wife, Christine Peterson, moved to a house in Los Altos, California. They were then president and secretary-treasurer, respectively, of the Foresight Institute, which they'd set up in 1987 to keep researchers current in the nano field. That fall, the institute held the first Foresight Conference on Nanotechnology, hosted by the Stanford University Department of Computer Science. This was a success, drawing 150 researchers from places like Du Pont, AT&T Bell Labs, IBM, MIT, Yale, and so forth. Many of these institutions had established their own laboratories for nanotechnology research. The consensus at the conference was that Drexler's intellectual creation had finally arrived, and that it was here to stay.

Ralph Merkle was there, of course, and when *Scientific American* reported on the conference in its January 1990 issue, it noted that Merkle "proposed that nanomachines be forbidden to have sex, thus preventing them from reshuffling their programs and surprising their creators."

Merkle had indeed said as much. "I plead guilty," he said. "It seems rather unnecessary to include sexual abilities in nanomachines, and would appear to require extra design effort to include them."

One of Merkle's friends, hearing about this, dubbed him a "nanopuritan."

In April 1990, two IBM researchers announced they had used a scanning tunnelling microscope to spell out the company's logo in thirty-five individual atoms of xenon. Richard Feynman, who had died two years before, would not have been surprised.

*O*n a cold February night in 1990, Hans Moravec debated Joe Weizenbaum at Villanova University, in Philadelphia. Weizenbaum, a professor at MIT and the author of a highly regarded book called *Computer Power and Human Reason,* had for a long time been meaning to review Moravec's *Mind Children* in a scholarly scientific journal but could never write more than a page or two before giving up in exasperation. How could you write about a plan for . . . *turning people into computers?* Weizenbaum never met Moravec until the day of the debate, but although he hated *Mind Children,* Weizenbaum found Moravec to be, as he put it, "a kind and gentle person." That didn't prevent him from launching into a long course of criticisms when debate time rolled around.

Moravec, Weizenbaum said, wanted to do away with the human gene pool. This ought to be regarded as a horrible prospect— morally equivalent to the Holocaust.

Moravec's desire for immortality through downloading was also a mistake: it would lead to uniformity and stagnation. The *good* thing about death, Weizenbaum said, was that it got rid of old and outmoded belief systems. It made room for variety, for the emergence of the new.

As for Moravec's *motive* behind his desire to create new and improved human beings—or to make robots that were even *better* than human beings—Weizenbaum had a speculation. Moravec was jealous of the way women were able to create new life: he was in fact plagued by a major case of . . . "uterus envy."

"I mean this quite seriously," Weizenbaum said.

But Moravec gave as good as he got. All he wanted to do, he said, was to hasten the evolution of human beings that was eventually going to occur anyway. What was wrong with that? Moravec wondered, in all innocence, why it was that Weizenbaum objected to continued human evolution only when it was done *intentionally.* Why wouldn't it be just as bad if it occurred naturally?

As for death being a good deal for the human race since it got rid of old and ossified belief structures, Moravec said that there was nothing in *Mind Children* that would prevent people from getting rid of their own mental contents. "People could erase parts of their memories if they wanted to," he said.

And as for "uterus envy," Moravec accepted the charge quite happily. "There's an attraction to building your own children," he said. "But this is not evil when *I* want to do it, any more than it's evil when a woman wants to do it."

*I*n 1986 Arthur C. Clarke was diagnosed as having amyotrophic lateral sclerosis (Lou Gehrig's disease). Later, in July 1988, he flew from Sri Lanka, where he lived, to Baltimore and entered Johns Hopkins Hospital, where he was given the new and much less serious diagnosis of post-polio syndrome.

Back in Sri Lanka, he was asked what he thought about becoming immortal either through cryonics or by means of Moravec's downloading methods.

"I must say I find Moravec's visions of the future pretty terrifying," Clarke said. "But then, I'm an old-fashioned conservative.

"I've never quite made up my mind about cryonics, although at the moment, I'm helping Alcor in its battle against the California legislature, and have quite a mountain of literature on the subject.

"Actually, I suspect that the question of immortality for humans is meaningless, since nobody really lives for more than about ten years anyway—after that we're a different entity! In my case it now seems only a couple of years—I suspect I've got what the Austrians wittily call 'Waldheimer's Disease.'"

*I*n February 1990, a second artificial-life conference was held at the Santa Fe Institute, in New Mexico. Just about everyone who attended the first conference showed up for the second one, plus about a hundred new converts.

Things got off to a good start when, early on in the conference, some participants noticed a *Wall Street Journal* article reporting that an IBM Labs researcher surprised himself when his computer started producing brain waves. "A major computer simulation of the brain has unexpectedly produced electrical waves resembling those in the brain itself," the article said.

Was *this* artificial life? It seemed like it.

At the Computer Virus Panel, it was clear that some of the conference participants were taking the whole idea of simulated computer life-forms pretty seriously. The main question for the panel was whether computer viruses (which some referred to as "liveware") were genuine artificial life-forms and, if so, whether they had their own *rights to life*—even as they were trashing your valuable computer data. There was no agreement one way or the other, but it was obvious that this was an extremely touchy issue.

In another session, Dave Jefferson presented the results of some new artificial-life experiments that he and colleagues were doing at UCLA. The university now had its very own Connection Machine—a massively parallel supercomputer—and on it they were running a new program called "Genesys." The program started with random bit-strings, out of which evolved colonies of antlike organisms that could find their way through mazes. Jefferson had as many as 131,072 of these computerized ants running through the mazes simultaneously, with varying degrees of success, and he brought with him the wiring diagram of his champion trail-following ant, the one who had made it through the computer's "John Muir Trail" in record time. He projected the ant's wiring diagram up on the screen proudly, much as if he were showing pictures of the wife and family.

But these were only *simulations*. As for *actual practice*—i.e., getting robots to do things out there in the real world—well, apparently this was always going to be a problem. At one of the demonstration sessions—the nightly Show 'n' Tells where people exhibited their prize ponies—one of the participants showed a videotape of his fabulous new robot. Actually it was only *part* of a robot, specifically an arm. This arm was outfitted with a computer vision system and a wooden paddle, to which a balloon was attached by a piece of string. When the arm got going, it paddled the balloon up and down—but slowly, as if the whole process were taking place under water.

The robot, said its inventor, was "juggling" the balloon. Next on the inventor's own personal research-and-development agenda was getting the robot to juggle the balloon *without* the string attached.

After that, maybe the robot could juggle *two* balloons. Or *three*.

In 1988, Dave Criswell came up with a new idea. It wasn't quite on the scale of taking the sun apart, but it was ambitious enough in its own way. He proposed holding the 2008 Olympic Games in space.

Criswell first presented the concept at a meeting of the American Institute of Aeronautics and Astronautics; later, both *Omni* and the Smithsonian's Institution's *Air & Space* magazine ran stories about the scheme. The idea was to build a two-mile-wide space station up in orbit, a structure big enough to hold ten thousand people.

Criswell had everything figured out: once around the ring would equal a ten-kilometer run; new zero-gravity sports events could be developed. He even invented a type of aircraft—a swing-wing space plane—that could get sports fans up there and back for the price of a typical ocean crossing. The swing-wing space plane would be an oval-shaped flying wing: the entire aircraft would be able to rotate in flight so that at lower altitudes it would fly like a regular aircraft, with its long axis to the wind, whereas once it rose above the atmosphere it would swivel around and fly up to space like a rocket. He got patents on this design and hoped to sell the idea to one of the major aerospace companies.

Robert Helmick, president of the U.S. Olympic Committee, was actually quite impressed by the Olympics-in-Space concept. "It's a fantastic idea, very creative," he said. "The Olympic Games have a universal appeal throughout the world, and I think it would be great to hold them in space and for there to be some visible insignia up there that everyone could see."

*I*n 1989 Bob Forward invented a new kind of satellite, a "statite." The name was short for "static satellite," meaning that the spacecraft would not orbit the earth: it would be *stationary*, floating up above the North Pole, held up as if by magic. Only it wouldn't be magic, of course, not a Bob Forward design. It would be held up there by a bank of lightsails, which would ride on the pressure of sunlight. They'd be like kites, only held up by the sun instead of by wind.

Large parts of the Soviet Union, of course, couldn't make use of geostationary satellites in equatorial orbit, but a North Pole statite would be perfect for communications, television, and everything else. Arthur Clarke had invented the communications satellite back in the 1940s but had neglected to patent the idea. He later wrote a wry piece about all the money he'd lost: "How I Lost a Billion Dollars in My Spare Time."

Forward was having Art Dula, a Houston attorney, patent the statite idea so he could sell it to the Russians. "I'm not going to let happen to *me* what happened to Arthur Clarke," Bob Forward said.

*W*hen Frank Tipler had written *The Anthropic Cosmological Principle* with John Barrow, he hadn't gone so far as to say that human life would evolve so as to achieve powers that were specifically *godlike;* he said only that at the Omega Point—that being the final and ultimate state of the universe—life would have *gone everywhere, done all things,* and so on. But in 1989 he went the rest of the way, speaking in explicit terms that went, if possible, beyond mere hubris to whatever else it was that lay out there.

"Life becomes omnipotent at the instant the Omega Point is reached," Frank Tipler said. "Since by hypothesis the information stored becomes infinite at the Omega Point, it is reasonable to say that the Omega Point is omniscient." Furthermore, since it exists everywhere, he added, "the Omega Point is omnipresent."

When all this happened, Tipler said, life would have brought the entire physical universe to a state of self-awareness. It would have transformed a dead cosmos into a living, thinking entity.

And that would be the end.

Acknowledgments

My thanks to all those who cooperated in the writing of this book, many of whom were kind enough to endure repeated interviews, and to read and correct one or more drafts of the chapters in which they appear. In addition, I'd like to thank several others who provided moral support, inspiration, ideas, and sources: Mark Collier, Doug Colligan, David M. Evans, Gerald Feinberg, David R. Forrest, Martin Gardner, Charles Griswold, Fred Hapgood, Steven G. Krantz, Allen J. Lopp, Michael A. G. Michaud, Roger J. Musser, Robert Nozick, Jim Oberg, Christine Peterson, Stanley Schmidt, and Steven Weinberg.

I owe a great personal debt to my wife, Pamela Regis, for her long indulgence over the duration of this work, for several ideas and insights, many good discussions, and—not least—one chapter title.

Profoundest thanks, finally, to my editor, Bill Patrick, who got me to thinking about the project in the first place, provided fine counsel from beginning to end, and (as before) saved me from many egregious misjudgments and stupidities during the course of it. For his gentle and patient guidance through some especially bad moments, I cannot thank him enough.

Selected Sources

The Mania

Barrow, John D., and Frank J. Tipler. *The Anthropic Cosmological Principle*. Oxford and New York: Oxford University Press, 1988.

Kent v. Carrillo. Case No. 191277. Reply of Plaintiffs to Defendants' Opposition to Application for Preliminary Injunction: Appendix of Declarations. California Superior Court, County of Riverside, 1988.

Chapter 1. Truax

Commercial Space Report 7 (September 1983): 1; (October 1983): 4.

Dille, Robert C., ed. *The Collected Works of Buck Rogers in the 25th Century*. New York: Bonanza Books, 1970.

Stein, Kathleen. "A Rocket for the People." *Omni* (October 1979): 71.

Truax, R. C. "Annapolis Rocket Motor Development, 1936–38." In *First Steps Toward Space* (Smithsonian Annals of Flight 10), edited by Frederick C. Durant III and George S. James, 259–301. Washington, D.C.: Smithsonian Institution Press, 1974.

Truax, Robert C. "The Conquest of Death." Unpublished manuscript, n.d.

———. "Towards a United States of the World." Unpublished manuscript, n.d.

Chapter 2. Home on Lagrange

Baird, John C. *The Inner Limits of Outer Space*. Hanover, N. H.: University Press of New England, 1987.

Heinlein, Robert A. *Farmer in the Sky*. [1950] London: Victor Gollancz, 1975.

Heppenheimer, T. A. *Colonies in Space*. New York: Warner Books, 1977.

Heppenheimer, Thomas A. "Resources and Recollections of Space Colonization." In *Space Colonization: Technology and the Liberal Arts*. AIP Conference Proceedings 148. Edited by C. H. Holbrow, A. M. Russell, and G. F. Sutton. New York: American Institute of Physics, 1986.

Leary, Timothy, with Robert Anton Wilson and George A. Koopman. *Neuropolitics: The Sociobiology of Human Metamorphosis*. Los Angeles: Starseed/Peace Press, 1977.

Meinel, Carolyn. "Silver Apples." Unpublished manuscript, n.d.

Michaud, Michael A. G. "Pro-Space: Interviews with the Space Advocacy." Center for the Study of Foreign Affairs, Foreign Service Institute, United States Department of State, Arlington, Va., 1985.

———. *Reaching for the High Frontier: The American Pro-Space Movement, 1972–84*. New York: Praeger, 1986.

Morse, J. T., and A. H. Smith. "Exercise Capacity in a Population of Domestic Fowl: Effects of Selection and Training." *American Journal of Physiology* 222 (June 1972): 1380.

Oberg, James. "OTRAG." *Omni* (June 1981): 69.

O'Neill, Gerard K. "The Colonization of Space." *Physics Today* (September 1974): 32.

———. *The High Frontier: Human Colonies in Space*. New York: William Morrow, 1977.

Poole, Robert, Jr. "African Deception." *Reason* (July 1978): 16.

Regis, Edward, Jr. "The Exodus Institute." *Omni* (March 1987): 20.

"Rockets for Sale." Narrated by Garrick Utley. "NBC Magazine with David Brinkley." NBC Television Network, New York. 20 March 1981.

Smith, A. H. "Physiological Changes Associated with Long-Term Increases in Acceleration." In *Life Sciences and Space Research XIV*, edited by P. H. A. Sneath. Berlin: Akademie-Verlag, 1976.

Smith, A. H., and C. F. Kelly. "Biological Effects of Chronic Acceleration." *Naval Research Reviews* 18 (1965): 1.

———. "Influence of Chronic Acceleration upon Growth and Body Composition." *Annals of the New York Academy of Sciences* 110 (1963): 410.

Woodbridge, Richard G., III. "A Forgotten Pioneer." *Proceedings of the IEEE* 67 (July 1967): 1085.

Chapter 3. Heads Will Roll

Darwin, Mike. "And Clifford Simak Passes." *Cryonics* 9 (June 1988): 17.

———. "The Door into Nowhere." *Cryonics* 9 (June 1988): 16.

Ettinger, R. C. W. "The Penultimate Trump." *Startling Stories* 17 (March 1948): 104.

Ettinger, Robert C. W. *The Prospect of Immortality.* Privately published, 1962. New York: Doubleday, 1964. London and Edinburgh: Sidgwick and Jackson, 1965.

Federowicz, Michael, Hugh Hixon, and Jerry Leaf. "Postmortem Examination of Three Cryonic Suspension Patients." Riverside, Calif.: Alcor Life Extension Foundation, 1984.

Heinlein, Robert A. *The Door into Summer.* New York: Ballantine, 1957.

Nelson, Robert F., as told to Sandra Stanley. *We Froze the First Man.* New York: Dell, 1968.

Simak, Clifford. *Why Call Them Back from Heaven?* New York: Ace, 1967.

Suda, I., K. Kito, and C. Adachi. "Viability of Long Term Frozen Cat Brain *in Vitro.*" *Nature* 212 (15 October 1966): 268.

Szilard, Leo. "The Mark Gable Foundation." [1948] *The Voice of the Dolphin and Other Stories.* New York: Simon & Schuster, 1961.

Chapter 4. Omnipotence, Plenitude & Co.

Becker, R. S., J. A. Golovchenko, and B. S. Swartzentruber. "Atomic-Scale Surface Modifications Using a Tunnelling Microscope." *Nature* 325 (29 January 1987): 419.

Darwin, Michael. "The Anabolocyte: A Biological Approach to Repairing Cryoinjury." *Life Extension Magazine* (July/August 1977): 80.

Darwin, Mike. "Resuscitation: A Speculative Scenario for Recovery." *Cryonics* 9 (July 1988): 33.

Dewdney, A. K. "Nanotechnology: Wherein Molecular Computers Control Tiny Circulatory Submarines." *Scientific American* (January 1988): 100.

Dietrich, J. "Tiny Tale Gets Grand." *Engineering and Science* (January 1986): 24.

Drexler, K. Eric. *Engines of Creation.* New York: Anchor-Doubleday, 1987.

———. Interview. *Omni* (January 1989): 66.

———. "Molecular Engineering: An Approach to the Development of General Capabilities for Molecular Manipulation." *Proceedings of the National Academy of Sciences* 78 (September 1981): 5275.

———. "Molecular Machinery and Molecular Electronic Devices." In *Molecular Electronic Devices II*, edited by Forrest Carter. New York: Marcel Dekker, 1987.

———. "Nanomachinery: Atomically Precise Gears and Bearings." Proceedings of the IEEE Micro Robots and Teleoperators Workshop, Hyannis, Mass., November 1987.

———. "A Technology of Tiny Things: Nanotechnics and Civilization." *Whole Earth Review* (Spring 1987): 8.

Feynman, Richard. "There's Plenty of Room at the Bottom." *Engineering and Science* 23 (February 1960): 22; reprinted in *Miniaturization*, edited by H. D. Gilbert, 282–96. New York: Reinhold, 1961.

MacDonald, Anson [Robert A. Heinlein]. "Waldo." *Astounding Science Fiction* (August 1942). Reprinted in Robert A. Heinlein: *Three by Heinlein: The Puppet Masters, Waldo, and Magic, Inc.* (New York: Doubleday, 1965).

Merkle, Ralph. "Molecular Repair of the Brain." *Cryonics* 10 (October 1989): 21.

Schmidt, Stanley. "Great Oaks from Little Atoms." Editorial. *Analog Science Fiction/Science Fact* (November 1987): 4.

Stiegler, Marc. "The Gentle Seduction." *Analog Science Fiction/Science Fact* (April 1989): 10.

Wowk, Brian. "The Death of Death in Cryonics." *Cryonics* 9 (June 1988): 30–36.

Chapter 5. Postbiological Man

Clarke, Arthur C. *The City and the Stars*. New York: Signet, 1957.

———. *The Lion of Comarre & Against the Fall of Night*. New York: Harcourt, Brace & World, 1968.

Dyson, Freeman J. *Infinite in All Directions*. New York: Harper & Row, 1988.

Ettinger, R. C. W. *Man into Superman*. New York: St. Martin's Press, 1972.

———. "Moravec's Duplicates." *The Immortalist* 20 (May 1989): 24.

Fjermedal, Grant. *The Tomorrow Makers*. Redmond, Wash.: Tempus Books, 1988.

Fredericksen, Dick. "I Have a Pipedream." *A Word in Edgewise* 5–8 (1971).

Moravec, Hans. Interview. *Omni* (August 1989): 74.

———. *Mind Children: The Future of Robot and Human Intelligence.* Cambridge, Mass.: Harvard University Press, 1988.

Nozick, Robert. *Anarchy, State, and Utopia.* New York: Basic Books, 1974.

———. *The Examined Life: Philosophical Meditations.* New York: Simon & Schuster, 1989.

Pohl, Frederik. "Intimations of Immortality." *Playboy* (June 1964): 158.

Tipler, Frank J. "The Omega Point as *Eschaton:* Answers to Pannenberg's Questions for Scientists." *Zygon* 24 (June 1989): 217.

Truax, Robert C. "The Conquest of Death." Unpublished manuscript, n.d.

Chapter 6. The Artificial Life 4-H Show

Dawkins, Richard. *The Blind Watchmaker.* New York: W. W. Norton, 1987.

———. *The Selfish Gene.* Oxford: Oxford University Press, 1976.

Drexler, K. Eric. "Biological and Nanomechanical Systems: Contrasts in Evolutionary Capacity." In *Artificial Life,* edited by Christopher G. Langton, 501–519. Reading, Mass.: Addison-Wesley, 1989.

Friedländer, Saul, Gerald Holton, Leo Marx, and Eugene Skolnikoff, eds. *Apocalypse: End or Rebirth?* New York and London: Holmes & Meier, 1985.

Gleick, James. "Artificial Life: Can Computers Discern the Soul?" *New York Times,* 29 September 1987, 18.

Henson, H. Keith. "Memes, Meta-Memes, and Politics." Unpublished manuscript, 1988.

Jeon, K. W., I. J. Lorch, and J. F. Danielli. "Reassembly of Living Cells from Dissociated Components." *Science* 167 (20 March 1970): 1626.

Langton, Chris. "Toward Artificial Life." *Whole Earth Review* (Spring 1988): 74.

Langton, Christopher G., ed. *Artificial Life.* Proceedings of an Interdisciplinary Workshop on the Synthesis and Simulation of Living Systems Held September 1987 in Los Alamos, New Mexico. Santa Fe Institute Studies in the Sciences of Complexity. Reading, Mass.: Addison-Wesley, 1989.

Moore, Edward F. "Artificial Living Plants." *Scientific American* (October 1956): 118.

Moravec, Hans. "Human Culture: A Genetic Takeover Underway." In *Artificial Life,* edited by Christopher G. Langton, 167–199. Reading, Mass.: Addison-Wesley, 1989.

————. "Machines from Molecules." Review of *Engines of Creation*, by K. Eric Drexler. *Technology Review* (October 1986): 76.

Pohl, Frederik. "When Machines Surpass People." Review of *Mind Children*, by Hans Moravec. *New Scientist* (February 1989): 65.

Reynolds, Craig W. "Flocks, Herds, and Schools: A Distributed Behavioral Model." *Computer Graphics* 21 (July 1987): 25.

Sagan, Carl. "Life." *Encyclopaedia Britannica*. 1970.

Taylor, Charles E., David R. Jefferson, Scott R. Turner, and Seth Goldman. "RAM: Artificial Life for the Exploration of Complex Biological Systems." In *Artificial Life*, edited by Christopher G. Langton, 275–95. Reading, Mass.: Addison-Wesley, 1989.

Chapter 7. Hints for the Better Operation of the Universe

Brooks, Harvey. "Technology-Related Catastrophes: Myth and Reality." In *Apocalypse: End or Rebirth?* edited by Saul Friedländer, Gerald Holton, Leo Marx, and Eugene Skolnikoff. New York and London: Holmes & Meier, 1985.

Criswell, David R. "Solar System Industrialization: Implications for Interstellar Migrations." In *Interstellar Migration and the Human Experience*, edited by Ben R. Finney and Eric M. Jones, 50–87. Berkeley: University of California Press, 1985.

Davidson, Frank. *MACRO: Big Is Beautiful*. New York: William Morrow, 1983.

Dyson, Freeman J. "Search for Artificial Stellar Sources of Infrared Radiation." *Science* 131 (June 1960): 1667.

————. "The Search for Extraterrestrial Technology." In *Perspectives in Modern Physics: Essays in Honor of Hans A. Bethe*, edited by R. E. Marshak, 641–55. New York: John Wiley, 1966.

Finney, Ben R. *Hokule'a: The Way to Tahiti*. New York: Dodd, Mead, 1979.

————. "Voyaging Canoes and the Settlement of Polynesia." *Science* 196 (17 June 1977): 1277.

Finney, Ben R., and Eric M. Jones, eds. *Interstellar Migration and the Human Experience*. Berkeley: University of California Press, 1985.

Goeller, H. E., and Alvin M. Weinberg. "The Age of Substitutability." *Science* 191 (20 February 1976): 683.

Hart, Michael H. "Interstellar Migration, the Biological Revolution, and the Future of the Galaxy." In *Interstellar Migration and the Human Experience*, edited by Ben R. Finney and Eric M. Jones, 278–91. Berkeley: University of California Press, 1985.

Jones, Eric M. and Ben R. Finney. "Fastships and Nomads: Two Roads to the Stars." In *Interstellar Migration and the Human Experience*, edited by Ben R. Finney and Eric M. Jones, 88–103. Berkeley: University of California Press, 1985.

Ley, Willy. *Engineers' Dreams*. New York: Viking, 1954.

Marchetti, C. "10^{12}: A Check on the Earth Carrying Capacity for Man." *Energy* 4 (1979): 1107.

Marchetti, Cesare. "How to Solve the CO_2 Problem without Tears." 7th World Hydrogen Conference, Hydrogen Today. Moscow, September 1988.

Niven, Larry. "Bigger than Worlds." *A Hole in Space*. New York: Ballantine, 1974.

———. *Ringworld*. New York: Del Rey-Ballantine, 1970.

Oberg, James Edward. *New Earths: Transforming Other Planets for Humanity*. Harrisburg, Pa.: Stackpole, 1981.

Regis, Edward, Jr. "Comet Odyssey." *Omni* (June 1984): 54.

———. "The Moral Status of Multigenerational Interstellar Exploration." In *Interstellar Migration and the Human Experience*, edited by Ben R. Finney and Eric M. Jones, 248–59. Berkeley: University of California Press, 1985.

———. "Mother Sun." *Omni* (December 1983): 122.

Thompson, David. "Astropollution." *CoEvolution Quarterly* (Summer 1978): 35.

Wetherill, George W. "Apollo Objects." *Scientific American* (March 1979): 54.

Chapter 8. Death of the Impossible

Bradbury, Ray. "A Sound of Thunder." In *R is for Rocket*. New York: Bantam, 1962.

Carter, Brandon. "Global Structure of the Kerr Family of Gravitational Fields." *Physical Review* 174 (1969): 1559.

Clarke, Arthur C. *Profiles of the Future*. New York: Bantam, 1960.

Davis, Philip J., and David Park, eds. *No Way: The Nature of the Impossible*. New York: W. H. Freeman, 1987.

Feinberg, Gerald. "Physics and Life Prolongation." *Physics Today* (November 1966): 45.

Forward, Robert L. "Alternate Propulsion Energy Sources." AFRPL TR-83-039. Edwards Air Force Base, Calif.: Air Force Space Technology Center, 1983; reprinted in *Commercial Space Report* 13 (October–November–December, 1989).

————. "Antigravity." *Proc. IRE* 49 (September 1961): 142.

————. "Antiproton Annihilation Propulsion." *J. Propulsion* 1 (September–October 1985): 370.

————. "Beamed Power Propulsion to the Stars." Paper presented at AAAS symposium, Interstellar Communication and Travel. AAAS annual meeting, Philadelphia, May 1986.

————. *Dragon's Egg.* New York: Del Rey-Ballantine, 1980.

————. "?EMAN A NI S'TAHW." *Mirror Matter Newsletter* 2 (March 1987): 3.

————. "Far Out Physics." *Analog Science Fiction/Science Fact* (August 1975): 147.

————. "Flattening Spacetime near the Earth." *Physical Review D* 26 (15 August 1982): 735.

————. *Flight of the Dragonfly.* New York: Bean Books, 1984.

————. *Future Magic.* New York: Avon, 1988.

————. "Gravity Sensors and the Principle of Equivalence." *IEEE Transactions on Aerospace and Electronic Systems* AES-17 (July 1981): 511

————. "Guidelines to Antigravity." *American Journal of Physics* 31 (March 1963): 166.

————. "How to Build a Time Machine." *Omni* (May 1980): 92.

————. "Roundtrip Interstellar Travel Using Laser-Pushed Lightsails." *J. Spacecraft* 21 (March–April 1984): 187.

Forward, Robert L., and Eugene F. Mallove. "Bibliography of Interstellar Travel and Communication." *Journal of the British Interplanetary Society* 27 (December 1974): 921; 28 (March 1975): 191; 28 (June 1975): 405; plus updates, 29 (1976): 494; 31 (1978): 225.

Forward, Robert L., and Larry R. Miller. "Generation and Detection of Dynamic Gravitational-Gradient Fields." *Journal of Applied Physics* 38 (February 1967): 512.

Gödel, Kurt. "An Example of a New Type of Cosmological Solutions of Einstein's Field Equations of Gravitation." *Reviews of Modern Physics* 21 (July 1949): 447.

Herbert, Nick. *Faster than Light: Superluminal Loopholes in Physics.* New York: New American Library, 1988.

Marshall, Eliot. "Garwin and Weber's Waves." *Science* 212 (15 May 1981): 765.

Plachta, Dannie. "The Man from When." *Worlds of IF* (July 1966); reprinted in Wollheim, Donald A. and Terry Carr, eds., *World's Best Science Fiction.* New York: Ace, 1967.

Rothman, Milton A. *A Physicist's Guide to Skepticism: Applying the Laws of Physics to Faster-than-Light Travel, Psychic Phenomena, Telepathy, Time*

Travel, UFO's and Other Pseudoscientific Claims. Buffalo, N.Y.: Prometheus, 1988.

Tipler, Frank J. "The Omega Point as *Eschaton*: Answers to Pannenberg's Questions for Scientists." *Zygon* 24 (June 1989): 217.

———. "Rotating Cylinders and the Possibility of Global Causality Violation." *Physical Review D* 9 (1974): 2203.

Weber, J. "Gravitational Waves." Gloucester, Mass.: Gravity Research Foundation, 1959.

———. "Observation of the Thermal Fluctuations of a Gravitational-Wave Detector." *Physical Review Letters* 17 (12 December 1966): 1228.

Chapter 9. *Laissez le Bon Temps Rouler*

Crease, Robert P., and Charles C. Mann. *The Second Creation*. New York: Macmillan, 1986.

Dyson, Freeman J. "Time Without End: Physics and Biology in an Open Universe." *Reviews of Modern Physics* 51 (1979): 447.

Forward, Robert L. "Extracting Electrical Energy from the Vacuum by Cohesion of Charged Foliated Conductors." *Physical Review B* 30 (15 August 1984): 1700.

Frautschi, Steven. "Entropy in an Expanding Universe." *Science* 217 (13 August 1982): 593.

Henson, H. Keith. "The Far Edge Committee." *Space-Faring Gazette* 3 (October 1987): 1.

———. "MegaScale Engineering and Nanotechnology Will Make Us Healthy, Wealthy, Wise, and Will Prevent Boredom." *Space-Faring Gazette* 4 (February 1988): 7; (March 1988): 7; (April 1988): 5.

Epilogue

Carroll, Paul B. "Computer Simulation Produces Activity Resembling Brain's Electrical Waves." *Wall Street Journal* (7 February 1990): B5.

Criswell, David R. "Multiconfiguration Reusable Space Transportation System." U.S. Patent 4,834,324 (30 May 1989).

———. "Orbital City: Put the Olympics in Space." Paper presented at International Panel on Space Exploration, AIAA Convention, Los Angeles, 1988.

Forward, Robert L. "Light-Levitated Geostationary Cylindrical Orbits Using Perforated Light Sails." *Journal of the Astronautical Sciences* 32 (April–June 1984): 221.

———. "The Statite: A Non-Orbiting Spacecraft." AIAA 89–2546. Paper presented at AIAA/ASME/SAE/ASEE 25th Joint Propulsion Conference, Monterey, Calif., July 1989.

Jefferson, David. "The Genesys System: Evolution as a Major Theme in Artificial Life." Paper presented at Second Artificial Life Conference. Center for Nonlinear Studies, Santa Fe Institute, New Mexico, February 1990.

Tipler, Frank J. "The Omega Point as *Eschaton*: Answers to Pannenberg's Questions for Scientists." *Zygon* 24 (June 1989): 217.

"Truax Engineering Sells Test Rocket to Navy." *Commercial Space Report* 12 (November–December 1988): 1.

Grateful acknowledgment is made to the following for the permission to reprint previously copyrighted material:

Excerpts from Kathleen Stein, "A Rocket for the People," *Omni* (October 1979), copyright © 1979 by Kathleen Stein and reprinted with permission of Omni Publications International, Ltd.

Lyrics to "Reach for the Stars" reprinted by permission of Carolyn Meinel.

Excerpts from Mike Darwin, "The Door into Nowhere," *Cryonics* 9 (June 1988), copyright © 1988 by the Alcor Life Extension Foundation, Inc. Used by permission.

Poem "The Man in the Can" reprinted by permission of Robert C. W. Ettinger.

Excerpts from Richard P. Feynman, "There's Plenty of Room at the Bottom," *Engineering and Science* (February 1960), copyright © 1960 by Alumni Association, California Institute of Technology. Used by permission.

Excerpts from interview with Eric Drexler, *Omni* (January 1989), copyright © 1989 by Ed Regis and reprinted with permission of Omni Publications International, Ltd.

Excerpts from Brian Wowk, "The Death of Death in Cryonics," *Cryonics* 9 (June 1988), copyright © 1988 by the Alcor Life Extension Foundation, Inc. Used by permission.

Poem "The Elusive Frozen Head" reprinted by permission of Curtis Henderson.

Excerpts from interview with Hans Moravec, *Omni* (August 1989), copyright © 1989 by Ed Regis and reprinted with permission of Omni Publications International, Ltd.

Excerpts from Hans Moravec, *Mind Children: The Future of Robot and Human Intelligence.* Cambridge, Mass., and London, England: Harvard University Press, 1988. Copyright © 1988 by Hans Moravec.

Image of Bush Robot created by Hans Moravec and Mike Blackwell, used by permission.

Excerpt from Frederik Pohl's letter to Ed Regis reprinted by permission of Frederik Pohl.

Excerpts from Robert C. W. Ettinger, "Moravec's Duplicates," *The Immortalist* 20 (May 1989), copyright © 1989 by the Immortalist Society. Used by permission.

Excerpts from Carl Sagan, "Life," in *Encyclopaedia Britannica,* 14th edition (1970), 13: 1083–1083A, reprinted by permission.

Excerpts from Ed Regis and Tom Dworetzky, "Child of a Lesser God," *Omni* (October 1988), copyright © 1988 by Ed Regis and Tom Dworetzky and reprinted with permission of Omni Publications International, Ltd.

Excerpts from H. Keith Henson, "MegaScale Engineering and Nanotechnology Will Make Us Healthy, Wealthy, Wise, and Will Prevent Boredom," *Spacefaring Gazette* 4 (February, March, April 1988), copyright © 1988 by Spacefaring Gazette. Used by permission.

Excerpts from Freeman J. Dyson, "Time without End: Physics and Biology in an Open Universe," copyright © 1978 by Freeman J. Dyson. Used by permission.

Excerpt from Arthur C. Clarke's letter to Ed Regis reprinted by permission of Arthur C. Clarke.

Excerpts from Frank J. Tipler, "The Omega Point as *Eschaton*: Answers to Pannenberg's Questions for Scientists," *Zygon* 24 (June 1989), copyright © 1989 by the Joint Publication Board of *Zygon*. Used by permission.

Index